NONSENSE MUTATION CORRECTION IN HUMAN DISEASES

NONSENSE MUTATION CORRECTION IN HUMAN DISEASES

An Approach for Targeted Medicine

HANA BENHABILES
JIESHUANG JIA
FABRICE LEJEUNE

Mécanismes de la Tumorigenèse et Thérapies Ciblées –
M3T – UMR, Université de Lille, Lille, France

Centre National de la Recherche Scientifique (CNRS) –
UMR, Paris, France

Institut Pasteur de Lille, Lille, France

Amsterdam • Boston • Heidelberg • London
New York • Oxford • Paris • San Diego
San Francisco • Singapore • Sydney • Tokyo

Academic Press is an imprint of Elsevier

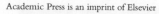

Academic Press is an imprint of Elsevier
125 London Wall, London EC2Y 5AS, UK
525 B Street, Suite 1800, San Diego, CA 92101-4495, USA
50 Hampshire Street, 5th Floor, Cambridge, MA 02139, USA
The Boulevard, Langford Lane, Kidlington, Oxford OX5 1GB, UK

British Library Cataloguing-in-Publication Data
A catalogue record for this book is available from the British Library

Library of Congress Cataloging-in-Publication Data
A catalog record for this book is available from the Library of Congress

ISBN: 978-0-12-804468-1

For information on all Academic Press publications
visit our website at http://store.elsevier.com/

Typeset by Thomson Digital

Publisher: Catherine Van Der Laan
Acquisition Editor: Jill Leonard
Editorial Project Manager: Fenton Coulthurst
Production Project Manager: Lucía Pérez
Designer: Maria Inês Cruz

CONTENTS

CONTENTS

ABOUT THE AUTHORS

LEJEUNE, FABRICE

Fabrice Lejeune is a researcher at the French Institute of Health and Medical Research (INSERM). In 2001, he received his PhD from Strasbourg University (France) after studying the alternative splicing of the splicing factor SRSF7 (9G8) in the laboratory of Dr James Stévenin at the Institute of Genetic and Molecular and Cellular Biology (IGBMC). From 2001 to 2005, he was postdoctoral fellow in the laboratory of Prof Lynne Maquat at the University of Rochester (USA) and studied NMD mechanism, mRNP composition on PTC-containing mRNA, and the pioneer round of translation. He then moved to Montpellier (France), to the Institute of Molecular Genetic of Montpellier (IGMM) in the laboratory of Prof Jamal Tazi, in order to start his research on the identification and characterization of molecules capable of rescuing the expression of genes harboring a nonsense mutation until 2008. He then started his laboratory at the Pasteur Institute of Lille until 2012 before joining the team of Dr David Tulasne who works on the mechanisms of apoptosis, signaling, and cancer at the Institute of Biology of Lille to go on the identification and the characterization of molecules capable of rescuing the expression of genes harboring a nonsense mutation.

JIA, JIESHUANG

Jieshuang Jia is a postdoctoral fellow studying NMD at the Institute of Biology of Lille (France) since Jun. 2015. Prior to her postdoctoral studies, she achieved her PhD at the University of Lille 2 by studying molecules with nonsense mutation correction capacity. She also received a Bachelor of Clinical Medicine from the Second Military Medical University in Jul. 2004 and a Master of Internal Medicine from the Nephrology Institute of Changzheng Hospital in Shanghai (China), with a specialty in Nephrology in Jun. 2007. She has won several first class awards and has been a merit student several times. She has worked in clinics for more than 4 years and received the Advanced Hospital Worker Status. During her time in clinic from Aug. 2007 to Oct. 2011, she directed students' internships and taught them diagnosis and treatment of diseases. She also performed some clinical studies on diseases such as polycystic kidney disease and chronic kidney disease with some connection to nonsense mutations, until 2011. Her interest focused on new diagnostic methods and therapies of diseases. She is now studying on NMD and would like to promote a strong interaction between nonsense mutation correction and clinic diseases.

BENHABILES, HANA

Hana Benhabiles is a PhD student working with Dr Fabrice Lejeune. She obtained her Master's degree in Genetics, in 2014, at the University of Lille (France). Prior to Lille, she did a Biological Engineering degree at Boumerdès University in Algeria. She acquired biomedical and biotechnological knowledge that she applies in her PhD. By coupling her engineering and research training, she is currently identifying new drugs correcting a nonsense mutation with a focus on their application in clinics in order to promote the expression of normally unexpressed proteins in these types of pathologies.

ACKNOWLEDGMENTS

The authors would like to thank the French national research agencies CNRS and Inserm, the Pasteur Institute of Lille, and the University of Lille. We are also undoubtedly thankful to patient associations like for instance but not limited to the Association française contre les myopathies, Vaincre la mucoviscidose, and l'Association pour la Recherche sur le Cancer in France but the same acknowledgment applied to patient associations of all countries in the world. Without the support and the help of these associations, we would not be able to develop our research programs. They are often at the origin of connections between research groups, academic and private laboratories, and a strong link maker between researchers, clinicians, and patients. Their energy is our engine to push away the limit of our knowledge. They have our deepest respect.

We also would like to thank our collaborators and in particular chemist colleagues with who the quest for new drugs would not have been possible. We deeply believe that interdisciplinary projects are the only efficient way for the development of new treatments. Finally, we are sincerely thankful to all our colleagues from the UMR8161 for the fruitful exchanges and their support, and also to all the colleagues of the fields of nonsense-mediated mRNA decay (NMD), premature termination codon recognition and suppression, and genetic disease area. These fields are lead by passionate researchers from who it is always an honor to receive guidance and advices. A long way is still needed to provide treatments for patients carrying nonsense mutations but thanks to the dynamism of the field, a real hope can be cultured to reach this final goal in a near future.

CHAPTER 1

General Aspects Related to Nonsense Mutations

Contents

DNA is the carrier molecule of the genetic information, and has to pass it on the daughter cells in respect of this information. Any modifications in the DNA molecule between two cell generations will result into a mutation. Besides maintaining the DNA molecule's integrity in order to preserve the genetic message, gene expression also has to reflect the encoded information carried by the DNA molecule and be delivered in an accurate way according to external and internal stimuli. To ensure this accuracy of the gene expression, quality controls are present for each step of gene expression from DNA replication until the folding and the posttranslational modifications of the protein (Araki and Nagata, 2011; Isken and Maquat, 2007; Kilchert and Vasiljeva, 2013; Liu et al., 2014; Lykke-Andersen and Bennett, 2014; Popp and Maquat, 2013; Porrua and Libri, 2015; Schmid and Jensen, 2013; Walters and Parker, 2014; Zhai and Xiang, 2014). This chapter will focus on one of these steps, that is, the mRNA quality control occurring after pre-mRNA splicing, and before the bulk of translation called nonsense-mediated mRNA decay (NMD), and on mechanisms directly related to it.

1 PREMATURE TERMINATION CODON, NONSENSE MUTATION, AND CONSEQUENCES ON GENE EXPRESSION

An open reading frame (ORF) starts with a translation initiation codon, which is often an AUG codon, and finishes with a stop codon (UGA, UAG, or UAA). When an additional stop codon is present inside of an ORF, meaning downstream of the initiation codon and upstream of the stop codon ending the encoding of the accurate C-terminal part of the wild-type protein, it is called premature termination codon (PTC). PTC can be introduced in an ORF, as a consequence of various events, such as a point mutation changing a coding codon into a stop codon (we then speak of "nonsense mutation"), or an insertion or a deletion inducing a frameshift mutation leading to the apparition of an in-phase PTC (Fig. 1.1 for the events at the DNA level). Insertions or deletions can occur at the DNA level by insertion or excision of DNA (transposable elements for example) or at the RNA level during pre-mRNA splicing, after a mutation located in an intron or in an exon and compromising the recognition of splice sites. Indeed, some mutations can induce partial or total intron retention, or total or partial exon skipping (Fig. 1.2). It is worth noting that all nonsense mutations are PTCs, but all PTCs are not nonsense mutations.

Statistical analysis of the distribution of the three stop codons at the normal translation termination position reveals that the UGA stop codon is the most frequent stop codon with 47% of the normal termination codons, then the stop codon UAA with 30% and, finally, UAG stop codon with 23% (Atkinson and Martin, 1994) (Table 1.1). Interestingly, the analysis of the distribution of the three stop codons as PTC is a little bit different, since the most frequent one is UGA, found in 51% of the nonsense mutations, then UAG with 31% and, finally, UAA with 18%. The frequency of the identity of nonsense mutations can be explained by the nature of codons that can be mutated into a stop codon. Indeed, TAG stop codon mainly comes from codons CAG (Gln) or TGG (Trp), TAA stop codon comes from mutations in codon CAA (Gln) or GAA (Glu), and TGA often derives from mutations in codon CGA (Arg) or TGG (Trp). This origin of stop codon can be partially explained by the fact that the most frequent mutation is the transition C→T (44%); this is induced by a deamination of the cytidine that converts the cytidine into a uracil which will be corrected into a thymidine (Fig. 1.3), since the mutation is occurring in the DNA molecule (Atkinson and Martin, 1994) (Table 1.1).

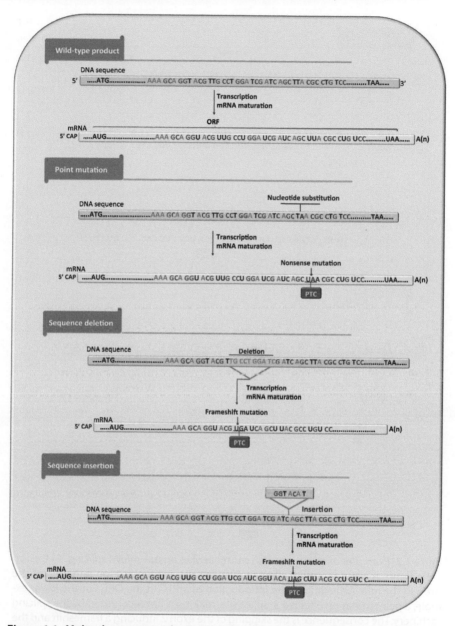

Figure 1.1 *Molecular events at the DNA level leading to the introduction of a PTC in the ORF of an mRNA.* The nucleotide triplet sequence is indicated and shows how various mutations can lead to the appearance of a PTC.

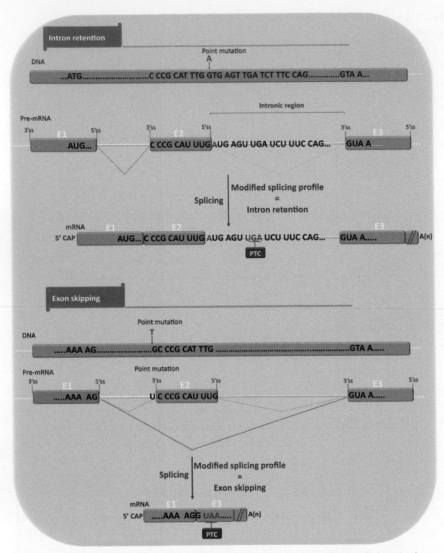

Figure 1.2 *Examples of mutations with consequences on pre-mRNA splicing.* At the top of the figure, a point mutation affects a 5′ splice site, promoting the retention of the intron in the mRNA, leading to the introduction of a PTC in the ORF. For the lower example, a point mutation affects a 3′ splice site, making it not recognized by the splicing machinery. The consequence is the skipping of the exon 2 inducing a frameshift and the introduction of a PTC in the ORF.

Table 1.1 Comparison of the frequency of each stop codon as physiological codons and as premature termination codons

Physiological STOP codon

Codon identity	Frequency (%)
UGA	47
UAG	23
UAA	30

Nonsense codon

Codon identity	Frequency (%)
UGA	51
UAG	31
UAA	18

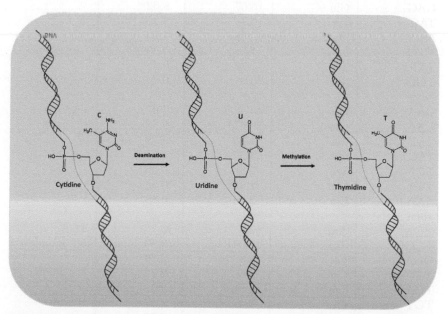

Figure 1.3 *Transition of a cytidine into a thymidine.* The cytidine is first subject to a deamination to generate a uridine. Since the reaction is occurring in DNA, the uridine is then subject to a methylation, in order to convert the uridine into a thymidine.

The statistical analysis of the mutated codons at the origin of a nonsense mutation shows that none of the codons encoding alanine, asparagine, aspactic acid, histidine, isoleucine, methionine, phenylalanine, proline, threonine, or valine can be replaced by a stop codon after a point mutation. UGA stop codon is the only stop codon to have exclusivity in the replacement of some amino acids. Indeed, codons encoding arginine, cysteine, or glycine can only be replaced by UGA stop codon after point mutation. In contrast, the three stop codons can be found replacing the position of a leucine or a serine (Table 1.2).

The consequence of the presence of a PTC on gene expression is rarely the synthesis of a truncated protein, as it will be explained later, but the silencing of the gene is due to the activation of a RNA surveillance mechanism called NMD that recognizes and degrades specifically mRNAs harboring a PTC.

Table 1.2 Identity and distribution of the codons leading to stop codon after a point mutation

Original codon	Associated amino acid	Per stop codon (%)	All stop codons (%)	Resulting STOP codon
CAG	Gln	41.07	12.84	TAG
TGG	Trp	23.21	7.26	
GAG	Glu	16.07	5.02	
TAC	Tyr	10.7	3.35	
AAG	Lys	5.35	1.67	
TTG	Leu	1.78	0.55	
TAT	Tyr	1.78	0.55	
TCG	Ser	0	0	
Total		100		
GAA	Glu	40	7.82	TAA
CAA	Gln	28.57	5.58	
TAT	Tyr	14.28	2.79	
TAC	Tyr	8.57	1.67	
TCA	Ser	2.85	0.55	
TTA	Leu	2.85	0.55	
AAA	Lys	2.85	0.55	
Total		100		
CGA	Arg	62.5	30.7	TGA
TGG	Trp	17.04	8.37	
GGA	Gly	5.68	2.79	
TGC	Cys	5.68	2.79	
TCA	Ser	4.54	2.23	
TGT	Cys	3.4	1.67	
TTA	Leu	11.3	0.55	
AGA	Arg	0	0	
Total		100		
Total			100	

Due to the degradation of the PTC-containing mRNA, the corresponding gene is not expressed at the protein level, leading to the silencing of the gene. NMD obeys specific rules and prevents the synthesis of truncated proteins with no function, with eventual harmful properties for the cell or, unfortunately, with partial or full wild-type activity (Bhuvanagiri et al., 2010; Karam et al., 2013; Kervestin and Jacobson, 2012; Popp and Maquat, 2014; Rebbapragada and Lykke-Andersen, 2009; Reznik and Lykke-Andersen, 2010; Schweingruber et al., 2013; Silva and Romao, 2009). The efficiency of NMD allows decreasing the level of PTC-containing mRNAs to 5–25% of the corresponding wild-type mRNA, meaning that a small proportion of PTC-containing mRNAs escape from NMD (Kuzmiak and Maquat, 2006). Mainly, such PTC-containing mRNAs are not translated as it was demonstrated by the absence of detectable protein synthesis from these mRNAs insuring the silencing of PTC-containing genes (You et al., 2007). However, rules have always exceptions: for instance, truncated HSP110 or p53 proteins synthesized from PTC-containing mRNAs have been reported (Dorard et al., 2011; Anczukow et al., 2008).

Various events can lead to the introduction of a PTC in a specific mRNA. Some of them are rare, such as errors leading to either generating a nonsense mutation or a frameshift mutation by insertion or deletion during DNA replication or transcription. However, the main sources of PTC come from splicing events and programmed DNA rearrangements occurring at specific loci, such as the T-cell receptor or the immunoglobulin genes (Fig. 1.4) (Delpy et al., 2004; Green et al., 2003; Wang et al., 2002).

Since pre-mRNA splicing events are the major source generating PTCs and because of the strong links between pre-mRNA splicing and NMD (see Section 3.3), a description of this mechanism might be helpful and can facilitate the understanding of the processing that leads to the identification of PTC during NMD.

2 PRE-mRNA SPLICING MECHANISM

2.1 Generalities

Pre-mRNA splicing is a general maturation process in higher eukaryotes, since only 700 genes are intronless, out of the 20,000–25,000 human genes; that is, about 3% of human genes (Busch and Hertel, 2013; Lander et al., 2001; Louhichi et al., 2011). Histone, interferon, or 50% of G-protein-coupled receptor genes constitute the major examples of intronless genes (Louhichi et al., 2011; Markovic and Challiss, 2009; Shabalina et al., 2010). Among the spliced pre-mRNAs, about 95% are also subject to alternative splicing, increasing the diversity of protein isoforms generated

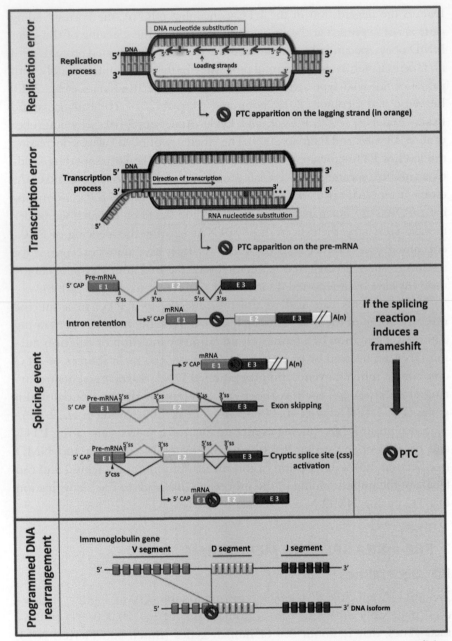

Figure 1.4 *Different sources of PTCs by replication or transcription errors, splicing events, or programmed DNA rearrangements.*

from one pre-mRNA and suggesting the complex regulation occurring during mRNA maturation, in order to synthesize the accurate protein isoform at the right moment and/or place (see Section 2.3) (Pan et al., 2008).

The splicing mechanism removes intronic sequences from a pre–mRNA in order to generate an mRNA consisting in exons only (Fig. 1.5). It should be kept in mind that not all exonic sequences are coding sequences. For example, the 5′ and the 3′ untranslated region (UTR) are noncoding exonic sequences. Similarly, some intronic sequences can be coding sequences when they are retained in the mRNA. Indeed, according to the profile of alternative splicing, an intron can become an exon by intron retention and an exon can become an intron due to exon skipping (for the different categories of alternative splicing, see Fig. 1.8).

The splicing reaction is ensured by a multiprotein and multi-RNA complex, called spliceosome (Galej et al., 2014; Matera and Wang, 2014;

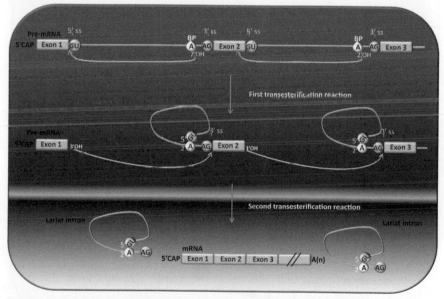

Figure 1.5 *Constitutive splicing reaction.* Two splicing reactions occurring on two consecutive introns. The splicing reaction starts with the nucleophile attack of the 2′ hydroxyl of the branch point on the exon/intron boundary in order to generate a free 5′ exon and a lariat intron linked to the 3′ exon by a first transesterification reaction. The second transesterification reaction is made of the nucleophile attack of the 3′ hydroxyl of the 5′ exon on the intron/exon boundary, in order to generate a free lariat intron and two linked exons. The splicing reaction occurs successively and/or simultaneously on all exons and introns of the pre-mRNA. Exons are represented by boxes and introns by gray lines. Splice sites are symbolized by circles containing "GU" for the 5′ splice site (5′ ss), "AG" for the 3′ splice site (3′ ss), or "A" for the branch point (BP).

Zhang et al., 2013). The core spliceosome is composed of 5 small nuclear ribonucleoproteins called U1, U2, U4, U5, and U6, and more than 150 proteins are involved in the recognition of splice sites (Wahl et al., 2009). All these components are sequentially loaded on the pre-mRNA, in order to play a role at a specific time in the splicing reaction (Fig. 1.6) (Chiou and Lynch, 2014).

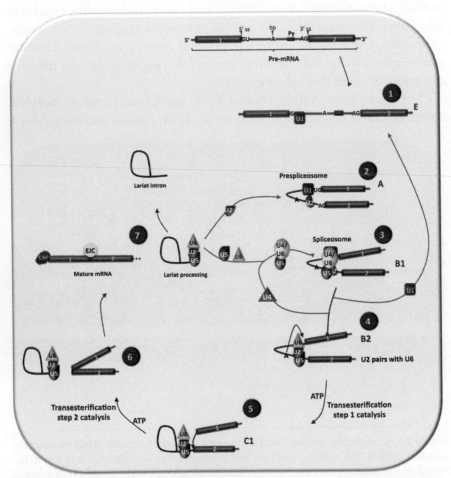

Figure 1.6 *Sequential loading of spliceosome components.* (1) U1 snRNP (U1) binds to the 5′ splice site in the E complex. (2) U2 snRNP binds the 3′ splice site in the A complex. (3) The tri-snRNP U4/U6 and U5 are loaded in the B1 complex. (4) U4 snRNP is released and U2 snRNP interacts with U6 snRNP to form the B2 complex. It is in this complex that the 2 transesterification reactions occur (5 and 6) to generate the mRNA harboring EJCs (7). The UsnRNPs are then recycled. Long black boxes represent exons and thick black lines represent introns.

The challenge of the spliceosome is to identify exonic sequences and intronic sequences inside a polynucleotide sequence. For that, introns are bordered by splice sites derived from consensus sequences, which are degenerated in superior eukaryotes. The 5′ splice site follows more or less a "GU–RAGU" consensus sequence, where R stands for purine bases (adenosine or guanosine). The 3′ splice site consists in a polypyrimidine tract (cytosine or uracil), often ended by a CAG sequence before the start of the exon. A third splice element, called branch point, is often located around 20–50 nucleotides upstream of the 3′ splice site. This last element is highly degenerated, and the only base shared by almost all branch points is an adenosine that will initiate the first transesterification reaction (Fig. 1.5). The strength of a splice site dictates its probability to be recognized by the spliceosome, and is related to its closeness to the consensus splice site sequences. A strong splice site would be constitutively recognized by the spliceosome. In contrast, the more the splice site sequence wanders from the consensus splice site sequence, the weaker is the splice site, meaning there are less chances to be recognized by the spliceosome. The notion of strength for the splice site is indeed crucial to keep in mind when the therapeutic approach wants to modulate splicing in regard to the exon skipping strategy (see Chapter C3).

Other elements influence the strength of a splice sites, and are called cis- and trans-acting regulators. Cis-acting regulators can be divided into enhancers or silencers of splicing and, since they can be located in exons or in introns, they will be named exonic splicing enhancers (ESE), intronic splicing enhancers (ISE), exonic splicing silencers (ESS), or intronic splicing silencers (ISS) (Fig. 1.7). These cis elements are bound by trans-acting factors that either activate or inhibit splicing. SR proteins and heterogeneous nuclear ribonucleoproteins

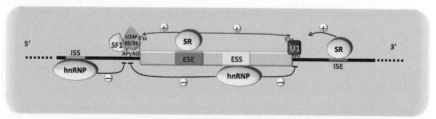

Figure 1.7 *Schematic representation of the* **cis-** *and* **trans-***splicing regulators.* Introns are represented by a thick black line and the exon is symbolized by a rectangle. SR proteins are noted by "SR." U1 snRNP (U1), splicing factor 1, and U2 auxiliary factor (U2AF) of 65 and 35 kDa are shown at the 5′ splice site (ss) and 3′ ss, respectively. Intronic splicing silencer (ISS), exonic splicing enhancer (ESE), exonic splicing silencer (ESS), and intronic splicing enhancer (ISE) are mentioned. On this model, SR proteins play the role of activators of splicing (+) when hnRNP are inhibitors of splicing (–).

(hnRNP) are two families of proteins that often antagonize for splicing activity (Caceres et al., 1994; Han et al., 2005; Zahler et al., 2004). The family of SR proteins consists of 12 members, named from SRSF1 to SRSF12, previously called ASF/SF2, SC35, SRP20, SRp75, SRp40, SRp55, 9G8, SRp46, SRp30c, SRp38, SRp54, and SRrp35, respectively (Fu, 1995; Manley and Krainer, 2010). SR protein members share structural and functional features, such as the presence of an RNA binding domain at the N-terminal part of the protein, an arginine/serine (RS) rich domain at the C-terminal part of the protein, and they are splicing factors with a positive or negative effect on splicing, depending on their binding sites (Bourgeois et al., 2004; Zhou and Fu, 2013). The second family often involved in the regulation of splice site use is called hnRNP. This family is composed of at least 20 members, named from hnRNP A to hnRNP U (Dreyfuss et al., 1993; Pinol-Roma et al., 1988). These proteins contain one or more RNA binding domains, a protein/protein interaction domain, a cellular localization domain, and a functional domain. At the origin, such proteins have been identified as proteins interacting with the pre-mRNA, also called heterogeneous nuclear RNA (hnRNA), explaining why they are called hnRNPs. They are involved in many different functions, from transcription to translation. They are often categorized as splicing inhibitors, even though a positive effect on splicing has been reported for some members of the hnRNP family, such as hnRNP G–T (Hui et al., 2003; Hung et al., 2008; Liu et al., 2009).

Interestingly, the strength of a splice site is actually dictated by the combination of the sequence of the splice site itself, the presence of *cis*-activator or inhibitor elements, and the proteins that recognize these *cis* elements. Indeed, a consensus splice site sequence can become a weak splice site if *cis*- or *trans*-inhibitor elements are located in the vicinity, and regulate it. In contrast, a splice site sequence different from the consensus sequence can be a strong splice site, if positive *cis* and *trans* elements regulate it. In addition, the strength of a splice site can be modulated according to the cell-type, the tissue, or various stimuli, meaning that, according to the physiological condition, a splice site can be recognized or ignored by the spliceosome (Coelho and Smith, 2014; Fu and Ares, 2014; Lee and Rio, 2015).

2.2 Categories of Alternative Splicing

Alternative splicing affects more than 90% of multiexonic pre-mRNAs in human (Blencowe, 2006; Johnson et al., 2003; Wang et al., 2008), demonstrating the almost universality of this process among human genes. Thanks to this mechanism, one gene can encode several proteins with various functions, tissue specificity and/or different regulation, explaining how the limited number of genes in human (about 25,000) can take charge of all the necessary cellular functions, estimated at 90,000 (Roy et al., 2013; Venter et al., 2001). Interestingly, the kinetic of removing introns subject to alternative splicing is slower than the kinetic to remove a constitutive intron, suggesting an additional level of the regulation for the expression of genes (Khodor et al., 2012; Pandya-Jones et al., 2013; Pandya-Jones and Black, 2009; Vargas et al., 2011). Such parameter could be interesting to keep in mind when a therapeutic strategy involving an action on splicing is designed.

Different categories of alternative splicing have been described (Fig. 1.8) (Roy et al., 2013; Wagner and Berglund, 2014). The most frequent alternative splicing event in human is the exon skipping, consisting in the no recognition

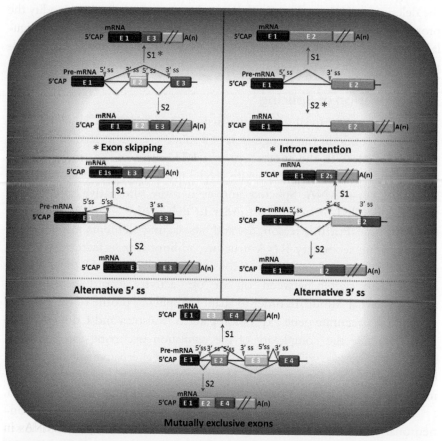

Figure 1.8 *The five main alternative splicing categories.* Exons are represented by a box and introns by a black horizontal line. For each category, the different splicing reactions are symbolized by a red line. The red star indicates the result of alternative splicing for exon skipping and intron retention. For the alternative 5′ ss or 3′ ss, the use of the upstream 5′ ss or the downstream 3′ ss, generates a shorter exon 1 or exon 2, indicated as E1s or E2s, respectively.

of the 3′ and 5′ splice sites surrounding an exon, and one of the less frequent alternative splicing categories is the intron retention, which is the no recognition and use of the 5′ and 3′ splice site of an intron by the spliceosome (Sammeth et al., 2008; Vitting-Seerup et al., 2014). Other alternative splicing categories include the competition between several 5′ or 3′ splice sites for the splicing of the same pre-mRNA region. The last category of alternative splicing is a competition between two exons to be included in an mRNA. This alternative splicing event, called mutually exclusive exons, induces the exclusion of one exon from the mRNA, when the competitor exon is included. This last category is the rarest alternative splicing category found in human (Vitting-Seerup et al., 2014). The consequences of an alternative splicing are the deletion or the insertion of a nucleic acid sequence that might modify the protein sequence encoded by the gene. Alternative splicing is used in particular to introduce or remove a protein domain or a regulator element, allowing the change of the corresponding protein or a modification in its expression.

2.3 Regulation of Splicing

Splicing is a very crucial step in the maturation of mRNA, since the shift of one nucleotide in the recognition of a splice site might lead to a frameshift mutation, with the appearance of a PTC most of the times. We saw that alternative splicing can generate several proteins with various functions from one pre-mRNA. Splicing, and even more alternative splicing, are tightly controlled processings in order to deliver the accurate mRNA at the right time. For this reason, *cis-* and *trans-*acting elements play a crucial role involving secondary RNA structure, additional *trans* factors, or a specific combination of *trans* factors (House and Lynch, 2008). All these regulatory elements switch between "on" or "off" activity, according to internal and/or external stimuli, in order to adapt the splicing profile of the cell, and generate the accurate gene expression pattern (Kalsotra and Cooper, 2011).

To illustrate the complexity of the regulation occurring for alternative splicing reactions, we will introduce one example of alternative splicing that has been deeply studied for many years. This example is about the alternative splicing occurring on the exon N1 of the cellular SRC kinase (c-src) pre-mRNA.

C-src gene encodes a tyrosine kinase expressed in neuronal and nonneuronal cells. Between the exon 3 and the exon 4, an 18-nucleotide sequence is recognized as an exon, called exon N1, only in nervous cells. The exon N1 codes for an SH3-type domain that affects the protein–protein interaction capacity of the c-src factor. This tissue-specificity involves *cis-*acting elements, upstream and downstream of the exon N1 (Fig. 1.9),

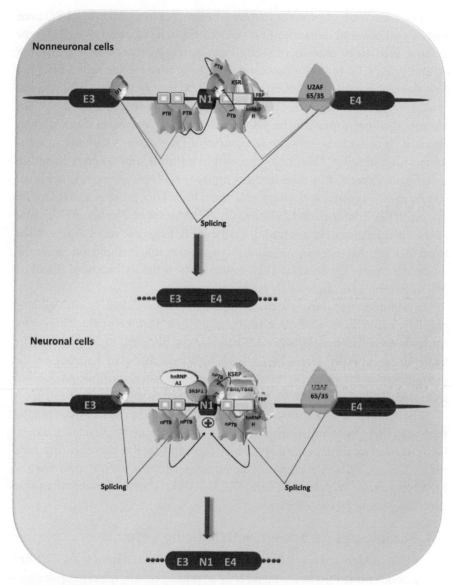

Figure 1.9 *Alternative splicing regulation of the exon N1 in nonneuronal (upper panel) or in neuronal cells (lower panel).* Exons are represented by dark blue cylinders and introns are symbolized by a thick dark blue line. The exons 3 (E3), N1 (N1), and 4 (E4) only from c-src pre-mRNA are represented. U1 snRNP is symbolized by a bicolor form harboring "U1." Other proteins are mentioned with their name on each model. The DCS is symbolized by a light blue rectangle. The binding sequences for PTB or nPTB are represented by an orange box.

and a combination of proteins involved in the splicing of the exon N1, and with a composition depending on the cell type. In all cell-types, KH-type splicing regulatory protein (KSRP), Fuse binding protein (FBP), hnRNP H, and hnRNP F bind the downstream regulator sequence called DCS for downstream control sequence (Black, 1992; Chan and Black, 1995; Chou et al., 1999; Min et al., 1995, 1997). The polypyrimidine tract binding (PTB) protein also interacts with the DCS, and with the upstream *cis*-regulatory element, in order to repress the recognition of the exon N1 by the spliceosome. Interestingly, PTB is not expressed in neurons, and another protein is specifically expressed in neurons, and replaces PTB by interacting with the upstream *cis*-regulatory element and the DCS. This protein, called nPTB (neural PTB) (Ashiya and Grabowski, 1997; Markovtsov et al., 2000) binds to the *cis*-regulatory elements and, unlike PTB, helps the recognition of the exon N1 by the spliceosome, and in particular the recognition of the 5′ splice site by the U1 snRNP. The consequence is the inclusion of the exon N1 in neuron cells, and not in other cell-types.

In addition to requiring intronic splicing elements, some exonic splicing regulators have been identified in the exon N1 (Rooke et al., 2003). This study showed that the exon N1 is bound by the SR protein SRSF1 and the hnRNP A1, hnRNP F, and hnRNP H. Interestingly, SRSF1 stimulates the recognition of the exon N1 by the spliceosome, when hnRNP A1 inhibits the splicing of N1.

The model of c-src shows the complexity of the alternative splicing regulation by involving intronic and exonic splicing regulators that can be activators or inhibitors of splicing, according to the proteins that bind them. This model also exemplifies the antagonistic effect of SR proteins and hnRNP factors (in particular hnRNP A1) and the involvement of a tissue specific splicing factor (nPTB).

2.4 Pathologies Associated with Splicing Defaults

Diseases related to a splicing default can occur by two different ways. The first one is about mutations affecting directly a splicing factor, which impairs specific or general splicing reactions. Often, such mutation is lethal, illustrating the dependence of cells for splicing. Indeed, splicing factors U2AF35 or SF3b subunit 4 have been shown to be essential during the early development stages of the zebrafish, after a screening using insertional mutagenesis inactivating these genes (Golling et al., 2002). Although that situation is rare, some pathologies have been shown to be related with the level of expression of splicing factors. For example, the SR protein SRSF3 (SRp20) has been shown to be overexpressed in different colorectal cancer cell lines (SW480, HT29, or DLD1) (Goncalves et al., 2008), and in many cancers (He et al., 2004, 2011; Jia et al., 2010). In contrast, downregulation of SRSF3 slows down the cell growth (He et al., 2011). The consequence of the modulation of the SRSF3 expression is a modification of the cellular splicing pattern. Among the targets of SRSF3, the pre-mRNA encoding the transmembrane receptor CD44 can be found (Galiana-Arnoux et al., 2003). Indeed, this transcript is subject to alternative splicing on several variable exons, named from v2 to v10, located between exon 5 and exon 16, as well as the exon 18 and the exon 19 that are mutually exclusive. Alternative splicing on this transcript generates RNA isoforms with different properties and, in particular according to the variable included exon, CD44 gains some metastatic properties, such as the isoforms with v8 to v10 (Yae et al., 2012). Interestingly, ESE have been found in these exons, and respond to SR proteins SRSF1, SRSF3, and SRSF7, leading to a retention of these exons in CD44 mRNA, when the level of these SR proteins increase (Galiana-Arnoux et al., 2003; Goncalves et al., 2008). Clearly, the level of some SR proteins influences, at least partially, the process of tumorigenesis by promoting the synthesis of particular CD44 mRNA isoforms.

SRSF3 is not the only splicing factor to have oncogenic property, since it has also been demonstrated for SRSF1, SRSF6, SRSF9, hnRNP A2/B1, and hnRNP H (Cohen-Eliav et al., 2013; Fu et al., 2013; Golan-Gerstl et al., 2011; Karni et al., 2007; Lefave et al., 2011). In contrast, some splicing factors have been shown to behave as tumor suppressors, such as RBM5, RBM6, and RBM10 (Bechara et al., 2013). A balance between tumorigenic and apoptotic splicing factors is essential to maintain a global splicing profile of the healthy cell.

Another pathology has been shown to be related with mutations in the splicing machinery. Indeed, some patients with retinitis pigmentosa carry mutations in the splicing factors PRPF31/U4-61k or PRP8 (Boon et al., 2007; Vithana et al., 2001; Wilkie et al., 2008). Several mutations in the splicing factor PRPF31 component of the U4 snRNP have been reported to cause retinitis pigmentosa. In particular, the missense mutation A216P leads to stabilize the interaction between PRPF31 and PRPF6, a component of the U5 snRNP. This stabilization promotes an inhibition of splicing, likely by preventing the disassembling of U4 and U5 snRNPs, in order to be recycled for a new splicing reaction. In the case of mutations affecting PRP8 and found in retinitis pigmentosa, another mechanism is involved. PRP8 is a component of the U5 snRNP and interacts with Brr2, a component that joins U5 snRNP during the maturation of this latter. Mutations found in patients with retinitis pigmentosa are located in the highly conserved C-terminal part of PRP8. The consequence of those mutations is a loss of interaction between PRP8 and Brr2, leading to an accumulation of the immature form of the U5 snRNP, and then an inhibition of splicing (Boon et al., 2007; Pena et al., 2007).

Although few cases of pathologies have been described to involve mutations in the splicing machinery, such mutations are often thought to be lethal. In most of pathologies related with a splicing failure, mutations affect one splicing reaction on a specific pre-mRNA, rather than the functionality of a general splicing factor. Examples describing mutations affecting either a splice site or a regulatory element are numerous in the literature. One example to illustrate how a specific splicing default can be at the origin of

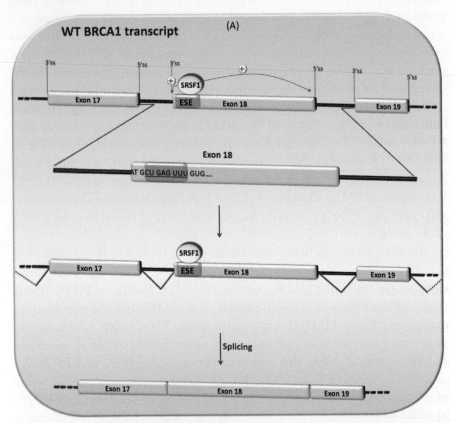

Figure 1.10 *Influence of a nonsense mutation in the exon 18 of the BRCA1 gene.* For the wild-type gene (A), SRSF1 interacts with an ESE to promote the splicing of introns 17 and 18 and to generate a wild-type BRCA1 mRNA, including the exon 18. When a G→T transversion mutation occurs

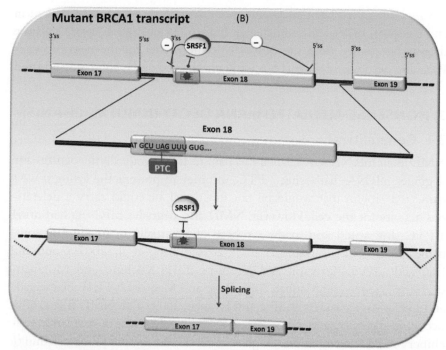

Figure 1.10 (cont.) (B), it generates a nonsense mutation and abolishes the recognition of the ESE by SRSF1, leading to the skipping of the exon 18 and a nonfunctional internally truncated BRCA1 protein.

a pathology concerns the BRCA1 gene. Mutations in the tumor suppressor gene BRCA1 are often found in hereditary breast and ovarian cancers (Miki et al., 1994). Among the mutations frequently found, the point mutation G6→T in the exon 18 changes a glutamic acid into a stop codon, and leads to the inactivation of an ESE bound by SRSF1 (Liu et al., 2001) (Fig. 1.10). Such mutation has two consequences: the first is the skipping of the exon 18, due to the inactivation of the ESE, generating an internally truncated BRCA1 missing 26 amino acids (D192→F1717) (Mazoyer et al., 1998). These 26 amino acids belong to a highly conserved region of the C-terminal domain of BRCA1 (BRCT), likely impairing the function of the BRCA1 protein, since this domain is involved in protein interactions (Yu et al., 2003). The second consequence is the decay by NMD of the PTC-containing BRCA1 mRNA isoform. In any case, the wild-type BRCA1 protein is not produced, leading to a higher susceptibility for the development of cancers.

After the mechanism of splicing, we can now see in detail the recognition of PTCs and the nonsense-mediated mRNA decay, in particular. Although this mechanism has been discovered more than 35 years ago in yeast (Losson and Lacroute, 1979), and soon after in human (Maquat et al., 1981), and was thoroughly studied from the early 1990s, the model for the recognition of PTC and the decay of the mRNA is still unachieved, and is constantly updated.

3 NONSENSE-MEDIATED mRNA DECAY (NMD) MECHANISM

3.1 Generalities

NMD is an mRNA surveillance mechanism that specifically recognizes and degrades mRNAs harboring a PTC, in order to prevent the synthesis of a truncated protein that would be not functional, or could carry a deleterious activity for the cell. However, NMD also degrades mRNAs harboring a PTC that would lead to the synthesis of a partially or totally functional truncated protein (Bhuvanagiri et al., 2010; Chang et al., 2007; Conti and Izaurralde, 2005; Kervestin and Jacobson, 2012; Kuchino and Muramatsu, 1996; Lejeune and Maquat, 2005; Popp and Maquat, 2014; Rebbapragada and Lykke-Andersen, 2009; Silva and Romao, 2009). The NMD mechanism occurs in the cytoplasm, consistent with the involvement of translation machinery in PTC recognition (Singh et al., 2007) during the pioneer round of translation or first round of translation (Ishigaki et al., 2001) and the following rounds of translation (Durand and Lykke-Andersen, 2013; Rufener and Mühlemann, 2013). Since, the mechanism of NMD can be different according to the species, we will describe here how NMD occurs in human cells.

The first round of translation occurs on a specific mRNP carrying CBP80 and CBP20 on the cap structure at the 5′ end of the mRNA, the poly(A) binding protein C1 (PABPC1) and N1 on the poly(A) tail at the 3′ end of the mRNA, and EJCs distributed upstream of exon–exon junctions (Chiu et al., 2004; Hosoda et al., 2006; Ishigaki et al., 2001; Lejeune et al., 2002, 2004; Sato et al., 2008). In the other rounds of translation that support the bulk of translation, CBP80/20 is replaced by eIF4E, PABPC1 is the only PABP present on the poly(A) tail, and no EJCs are present any longer on the mRNA (Fig. 1.11).

In mammalian cells, two models attempt to explain how NMD can distinguish between PTCs and normal termination codons. The historically first model proposes that a PTC is defined as a translation termination codon, located more than 50–55 nucleotides upstream of at least one exon–exon junction. This model suggests a strong link between pre-mRNA splicing and NMD. Indeed, a protein complex called exon junction complex (EJC) is deposited 20–24 nucleotides upstream of all exon–exon junctions, as a consequence of

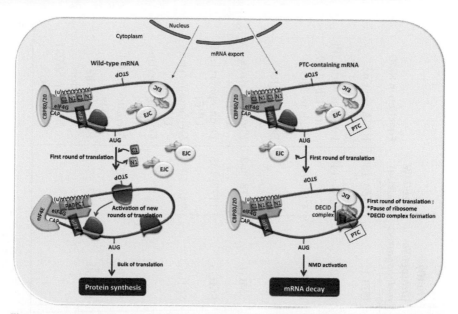

Figure 1.11 *Consequences of the first (pioneer) round of translation occurring on wild-type mRNA and on PTC-containing mRNA.* The first round of translation occurs on mRNA carrying EJCs, CBP80/CBP20 on the 5′ end, and a mix of the poly(A) binding protein C1 (PABPC1; C1) and the poly(A) binding protein N1 (PABPN1; N1) on the poly(A) tail at the 3′ end. On a wild-type mRNA (left part), EJCs, CBP80/CBP20, and PABPN1 will be removed from the mRNA after the pioneer round of translation and eIF4E will replace CBP80/CBP20 on the CAP structure, in order to support the bulk of translation. On a PTC-containing mRNA (right part), the mRNA will still carry at least one EJC, CBP80/CBP20, and the PABPN1 PABPC1, in order to promote NMD.

splicing (Le Hir et al., 2000a,b, 2001; Le Hir and Seraphin, 2008). The second model proposes that NMD is activated when the distance between the stop codon and the cytoplasmic PABPC1 is abnormally long (Behm-Ansmant et al., 2007; Eberle et al., 2008; Fatscher et al., 2014; Silva et al., 2008; Singh et al., 2008). This model necessitates a molecular mechanism measuring this distance and determining whether the length is normal or not.

NMD requires dozen of central proteins in order to identify a PTC from a normal termination codon. These proteins are called Up frameshift (UPF) proteins, after their identification in the yeast *Saccharomyces cerevisiae* (Culbertson et al., 1980; Leeds et al., 1991, 1992), or suppressors with morphogenetic defaults of genitalia (SMG) named like that after their identification in *Caenorhabditis elegans* (Cali et al., 1999; Hodgkin et al., 1989) (see Table 1.3 for the correspondence). The exact rule of these proteins in the mechanism of NMD still remains unclear, even though all these proteins

Table 1.3 NMD factors present in mammals, yeast, worm, and fruitfly

NMD factors (mammals)	Size/MW (aa/kDa)	Yeast (S. cerevisiae)	Size/MW (aa/kDa)	Worm (C. elegans)	Size/MW (aa/kDa)	Fruitfly (Drosophila melanogaster)	Size/MW (aa/kDa)	Function in NMD
UPF1 (Rent1)	1118/123	Upf1p	971/109	Smg-2	1069/120	UPF1	1180/130	RNA helicase
UPF2 (Rent2)	1272/148	UPF2p	1089/127	Smg-3	1142/132	UPF2	1241/140	Interacts with UPF1 and UPF3/3X
UPF3 (UPF3a)	452/52	UPF3p	387/45	Smg-4	329/39; 368/43	UPF3	492/56	Binds to EJC
UPF3X (UPF3b)	483/58							Binds to EJC
SMG1	3661/411			SMG1	3640/405	SMG1	3218/362	Phosphorylates UPF1
SMG5 (EST1B)	1016/114			SMG5	549/64	SMG5	1177/135	Dephosphorylates UPF1
SMG6 (EST1A)	1419/161			SMG6	1241/140	SMG6	948/109	Dephosphorylates UPF1
SMG7 (EST1C)	1137/127			SMG7	458/53			Dephosphorylates UPF1
SMG8	991/110			SMG8	873/100	SMG8	944/107	Mediates SMG1 kinase
SMG9	520/58			SMG9	385/44	SMG9	487/55	Promotes the interaction between SMG1 and SMG8

The size of amino acids (aa) and the molecular weight (MW) are indicated for each factor.

have been intensively studied and characterized in order to generate structural, cellular, and functional data that we are going to review.

3.2 Main proteins involved in NMD

3.2.1 UPF1/RENT1/SMG2

The UPF1 gene is carried by the chromosome 19 and encodes a 123 kDa protein with several functional domains, such as a CH-domain, a helicase domain, and an S/Q domain (Fig. 1.12). An ultraconserved domain is present at the N-terminal part of the protein and this domain has been recently shown to be cleaved during apoptosis by caspases and promoting apoptosis (Jia et al., 2015; Popp and Maquat, 2015). The role of the ATP-dependent helicase activity in NMD has been studied for a while, and it started with the discovery that a mutant version of UPF1 impairing the helicase domain (after the mutation of the arginine at the position 843 into a cysteine: R843C) leads to a dominant negative protein inhibiting NMD (Sun et al., 1998). It is only recently that the function of this domain in NMD has been clarified (see next paragraph).

UPF1 interacts with a wide net of proteins (Varsally and Brogna, 2012), so we will concentrate only on proteins with demonstrated role in NMD. Prior to being recruited to the EJC, UPF1 first interacts with the cap binding protein CBP80 via its helicase domain (Hwang et al., 2010). This interaction promotes the recruitment of another partner of UPF1, the protein kinase SMG1, associated with its partners, the proteins SMG8 and SMG9, via the SQ domain. In addition, prior to being recruited to the mRNA, UPF1 also interacts with the release factors 1 and 3, in a complex called

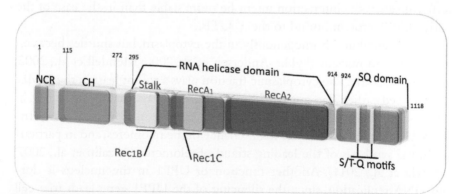

Figure 1.12 *Schematic representation of UPF1 and its functional domains.* Amino acid positions are indicated at the top. The N-terminal conserved region (NCR); the cysteine-histidine-rich (CH) domain; the helicase domain with the RecA1, RecA2, Rec1B, and Rec1C motifs; and the serine-glutamine-rich (SQ) domain, including the recognition and phosphorylation S/T-Q motifs are indicated.

SURF (for SMG1-UPF1-eRF1 and eRF3 complex) that works as a translation termination complex (Ivanov et al., 2008; Kashima et al., 2006). UPF2 and UPF3X then recruit UPF1 (in the SURF complex) to the EJC, via its CH domain at least, and will stimulate its helicase activity (Chakrabarti et al., 2011; Chamieh et al., 2008) in order to form the DECID (DECay InDucing) complex (Kashima et al., 2006). The interaction of UPF1 with UPF2 modifies the spatial conformation of UPF1 that becomes active for its 5′–3′ unwinding activity, thanks to its helicase domain. Interestingly, the helicase activity of UPF1 is tightly controlled since the N-terminal domain of UPF1 (CH domain), as well as the C-terminal domain (SQ domain), influences negatively the helicase activity. It is only when UPF1 engages interactions by its CH and SQ domains with multipartners that the helicase activity is released (Fiorini et al., 2013). Once the ATPase/helicase activity is unlocked, UPF1 will displace proteins bound to the downstream part of the PTC, in order to allow its fast degradation (Franks et al., 2010).

Recently, UPF1 has been demonstrated to bind mRNAs independently to the EJC, all over the ORF and on the 3′ UTR. Interestingly, when ribosomes read the ORF, they displace UPF1 present on the ORF resulting in a concentration of UPF1 in the 3′ UTR (Hogg and Goff, 2010; Kurosaki and Maquat, 2013; Zund et al., 2013). These data are in opposition to previous works showing that NMD is activated on an mRNA by tethering UPF proteins, such as UPF1, downstream of a physiological stop codon (Gehring et al., 2003; Gonzalez-Hilarion et al., 2012; Lykke-Andersen et al., 2000). However, its apparent opposition might be linked to the stability of the interaction between the UPF protein and the 3′ UTR: in the case of the tethering assay, the interaction might be more stable than in the case of the native UPF1 protein bound to the 3′ UTR.

UPF1 distributes homogenously in the cytoplasm, but shuttles between cytoplasm and nucleus (Lykke-Andersen et al., 2000; Mendell et al., 2002; Serin et al., 2001). The cytoplasmic fraction plays a role at least in the NMD, as described earlier. The nuclear fraction of UPF1 plays various functions, in addition of NMD (Imamachi et al., 2012; Varsally and Brogna, 2012). Indeed, UPF1 is involved in the maintenance of the telomeres, and in particular in the synthesis of the leading strand of telomeres (Azzalin et al., 2007; Chawla et al., 2011). Another function of UPF1 in the nucleus is during DNA replication, since the silencing of the UPF1 gene leads to a cell cycle arrest at the S-phase (Azzalin and Lingner, 2006b). Consistent with this function, UPF1 interacts with the DNA polymerase δ in an RNA-independent manner (Azzalin and Lingner, 2006b; Carastro et al., 2002).

This interaction raises the possibility that the helicase domain of UPF1 plays a role in the replication fork progression.

Based on the fact that UPF1 is predominantly cytoplasmic and only a small proportion of UPF1 resides in the nucleus, it is possible that the nuclear UPF1 has only a regulation role in DNA replication, telomere maintenance, or other yet unidentified nuclear functions. In addition, it is also a possibility that the nuclear UPF1 already plays a role in NMD by binding the 3′ UTR before other proteins cover that region of the mRNA, since UPF1 concentrates on 3′ UTR (Hogg and Goff, 2010; Kurosaki and Maquat, 2013; Zund et al., 2013). This binding of UPF1 and in particular the phosphoisoform of UPF1 in the 3′ UTR of mRNAs could play different functions. For instance, phospho-UPF1 has been shown to interact with the stem loop binding protein (SLBP) that binds to the stem loop present at the 3′ end of histone mRNAs. SLBP is responsible for the regulation of the translation and the stability of histone mRNAs. This responsibility could be achieved via its interaction with UPF1 that would next recruit SMG5 and proline-rich nuclear receptor coactivator 2 (PNRC2) protein in order to induce the mRNA decay (Fig. 1.13) (Choe et al., 2014).

3.2.2 UPF2/RENT2/SMG3

UPF2 is a predominantly cytoplasmic protein that concentrates around the nucleus (Lykke-Andersen et al., 2000; Serin et al., 2001). A nuclear localization sequence (NLS) is present in the N-terminal region of UPF2, suggesting that this protein can also go to the nucleus (Serin et al., 2001). Human UPF2 gene is located on the chromosome 10 and encodes for a protein of about 150 kDa. Several functional domains have been found in UPF2, and in particular three middle domains of eIF4G (MIF4G) in the N-terminal and in the central region of UPF2, and a UPF1 interacting domain in the C-terminal part (Fig. 1.14). UPF2 interacts with UPF1 and with UPF3/UPF3a or UPF3X/UPF3b in particular, to activate both the ATPase and the helicase function of UPF1 (Chamieh et al., 2008). Interestingly, UPF2 or UPF2 associated with UPF3X/UPF3b, binds RNA at least in vitro, but the functional relevance of that activity remains to be clarified (Kadlec et al., 2004). UPF2, like UPF1, is a phosphoprotein (Chiu et al., 2003) whose involvement in phosphorylation in crucial interactions with other NMD factors has been shown only in the yeast S. cerevisiae, but not yet in mammals (Wang et al., 2006).

The role of UPF2 in NMD remains incompletely characterized, according to the limited functions in the NMD mechanism of this protein identified up to the present time: interaction with UPF1 and UPF3/UPF3a or

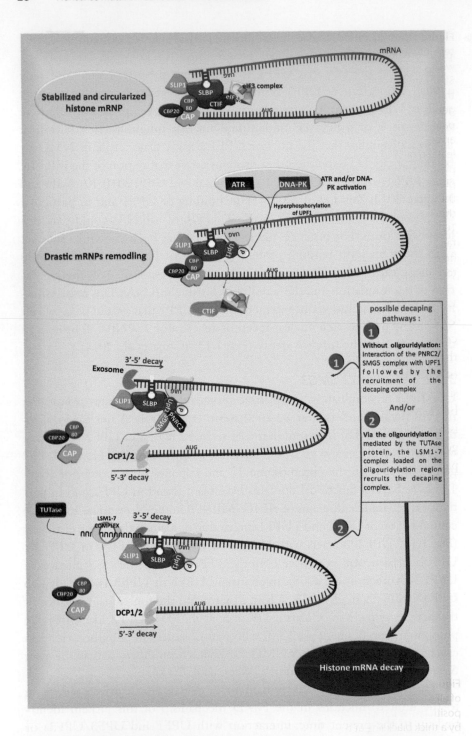

▶ **Figure 1.13** *Histone mRNA decay pathway.* During histone mRNA translation, SLBP protein interacts with a stem loop located in the 3′ UTR of the histone mRNA. SLBP interacts with CBP80 to circularize the mRNA and promote an efficient translation. In addition, SLBP interacts with SLBP interacting protein 1 (SLIP1) and the CBP80/CBP20-dependent translation initiation factor (CTIF). Under conditions requiring the silencing of histone genes, such as the inhibition of the DNA replication, UPF1 is recruited on the ribosome pausing on the physiological stop codon and becoming hyperphosphorylated by ATR and/or DNA-PK proteins. UPF1 hyperphosphorylated interacts with SLBP, inducing the leaving of CTIF. Then SMG5 and PNRC2 are recruited by UPF1 which induces the decapping of the 5′ end of the histone mRNA by helping in the recruitment of the decapping complex Dcp1/Dcp2 and the recruitment of 5′–3′ exoribonuclease activity. Due to the unprotected 3′ end of the mRNA, the exosome is loaded and degrades mRNA from the 3′ end to the 5′ end. An alternative and/or concomitant way is the recruitment of a TUTase by hyperphosphorylated UPF1, in order to induce the addition of a poly uridine tail at the 3′ end of the mRNA; this will recruit the Lsm1–7 complex. That complex then helps in the recruitment of the decapping enzymes Dcp1/Dcp2 at the 5′ end, and the 3′ to 5′ decay enzymes at the 3′ end of the histone mRNA.

UPF3X/UPF3b, and activation of the helicase activity of UPF1. In addition, UPF2 seems not to be absolutely required for some NMD reactions, according to the composition of the EJC (Gehring et al., 2005), suggesting that several different pathways can activate NMD. However, the involvement of UPF2 in NMD might be more predominant than originally thought, since it is one of the UPF proteins with UPF1, unlike UPF3X/UPF3b, to be cleaved by caspases during apoptosis (Jia et al., 2015). The fact that UPF2 is targeted by caspases in order to shut down NMD during apoptosis suggests a central role of that protein in NMD and/or an involvement in other pathways that has to be blocked to allow cell death progression. The UPF2

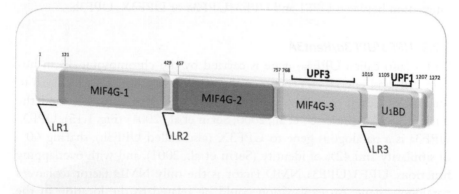

Figure 1.14 *Schematic representation of UPF2 functional domains.* The middle domain of eIF4G [MIF4G (1–3)] and UPF1 binding domain (U1BD) are represented. The amino acid positions are indicated at the top. The interacting surfaces with UPF3 and UPF1 are shown by a thick black line at the top. Finally, the linker domains 1–3 are shown (LR1–LR3).

caspase-cleavage fragments promote new activities, providing new information on the functional properties of the UPF2 domains. The N-terminal fragment, for instance, shows capacity at promoting apoptosis and at inhibiting NMD, when it is overexpressed in cells (Jia et al., 2015). The NMD inhibition by the N-terminal fragment could suggest that this part of UPF2 is involved in interactions with other NMD factors, but is unable to assume the function of UPF2 in NMD, making this fragment a dominant negative. In support of this statement, a highly conserved N-terminal region has been found in yeast UPF2, in which mutations impair with NMD (Fourati et al., 2014). The C-terminal part has been shown to interact with UPF1 and SMG1. This C-terminal part alone retains the function of UPF2 in NMD, explaining why an overexpression of this fragment does not interfere with NMD (Clerici et al., 2013; Jia et al., 2015; Kadlec et al., 2004). UPF2 was thought to be simply the bridge between UPF1 and UPF3/UPF3a or UPF3X/UPF3b bound to the EJC (Kashima et al., 2006). However, this model starts to be challenged, since UPF2 could stimulate the endonucleolytic cleavage activity of SMG6 in an EJC-independent way, meaning that UPF2 is not necessarily recruited to the EJC in order to function in NMD (Boehm et al., 2014). It also means that a bridge between UPF1 and UPF3/UPF3a or UPF3X/UPF3b is not absolutely required for NMD. Supporting this vision, electron microscopy study of complexes, including SMG1, SMG8, SMG9, UPF1, and UPF2, shows that UPF2 can be recruited to the complex formed by SMG1/8/9 and UPF1, without the presence of the EJC or UPF3/UPF3a or UPF3X/UPF3b (Melero et al., 2014). Overall, these data suggest that the role of UPF2 in NMD is not limited to promoting the interaction between UPF1 and UPF3/UPF3a or UPF3X/UPF3b.

3.2.3 UPF3/UPF3a/Rent3A

UPF3 (also called UPF3a) gene is carried by the chromosome 13 in human cells and encodes a protein of about 52 kDa that mainly localizes to the nucleus, even though it can shuttle between the cytoplasm and the nucleus (Lykke-Andersen et al., 2000; Serin et al., 2001) (Fig. 1.15). UPF3/UPF3a is a paralogous gene to UPF3X (also called UPF3b), sharing 60% of similarity and 42% of identity (Serin et al., 2001), and with overlapping functions. UPF3/UPF3a NMD factor is the only NMD factor to have a paralogous gene. An explanation could be related to the location of the UPF3X/UPF3b gene on the chromosome X. Indeed, many genes located on chromosome X have a paralog on an autosomal chromosome, since chromosome X is inactivated during spermatogenesis and some functions

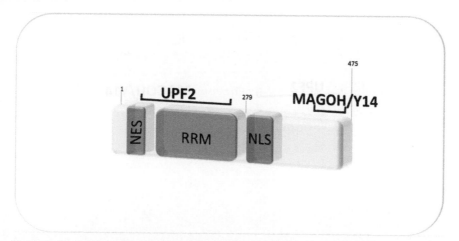

Figure 1.15 *Schematic representation of UPF3 and its functional domains.* Amino acid positions are indicated at the top. The NES, the RRM, and the NLS are shown. The interacting surfaces with UPF2 and MAGOH/Y14 are indicated by a thick black line at the top.

carried by the chromosome X are still needed during spermatogenesis (Wang, 2004). UPF3/UPF3a and UPF3X/UPF3b have been shown to interact with EJC components Y14, Ref/Aly, RNPS1, and SRm160 in an RNase insensitive way (Kim et al., 2001; Lejeune et al., 2002), and UPF1 and UPF2 (Chan et al., 2009; Lejeune et al., 2002, 2003). Although UPF3/UPF3a is less efficient than UPF3X/UPF3b in promoting NMD (Gehring et al., 2005; Lykke-Andersen et al., 2000), UPF3/UPF3a can compensate a decrease in the level of UPF3X/UPF3b (Chan et al., 2009). However, natural substrates of NMD are not indifferentially regulated by UPF3/UPF3a or UPF3X/UPF3b, since some mRNAs upregulated in the absence of UPF3X/UPF3b are not affected by a downregulation of UPF3/UPF3a, such as GADD45B, ATF3, ASNS, or MAFF genes (Chan et al., 2007). Due to the low involvement of UPF3/UPF3a in NMD, UPF3X/UPF3b has been more studied than UPF3/UPF3a in the recognition of PTC and in the NMD mechanism.

3.2.4 UPF3X/UPF3b/Rent3B

The UPF3X/UPF3b gene is located on the chromosome X and encodes a protein of about 58 kDa (Lykke-Andersen et al., 2000; Serin et al., 2001). UPF3/UPF3a and UPF3X/UPF3b are paralogs, and share the yeast UPF3 gene as their common ancestor gene. UPF3X/UPF3b is a mainly nuclear protein with the capacity to shuttle between the nucleus and the cytoplasm, thanks to the presence of four nuclear localization sequences (NLS) located

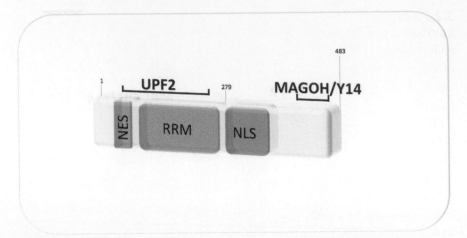

Figure 1.16 *Schematic representation of UPF3X/UPF3b and its functional domains.* The NES, NLS, and the RRM are represented. The interaction surfaces with UPF2 and MAGOH/Y14 are shown. The amino acid position is indicated at the top.

in the middle and in the C-terminal part of the protein, and one nuclear export sequence (NES) located at the N-terminal part of the protein. UPF3X/UPF3b has also several other clearly identified domains, such as an RRM (RNA recognition motif) and some interacting domains with UPF2 and MAGOH/Y14 (Fig. 1.16).

According to the current model, UPF3X/UPF3b is thought to be the first NMD factor to be loaded on the mRNA, before the export of the mRNA to the cytoplasm (Le Hir et al., 2001). The recruitment of UPF3X/UPF3b on the mRNA is operated by the EJC via its C-terminal domain interacting with three proteins of the core EJC (eIF4AIII, Y14, and MAGOH) (Buchwald et al., 2010; Gehring et al., 2003). UPF3X/UPF3b also interacts with UPF2 via its N-terminal domain. Although downregulation of UPF3X/UPF3b impairs many NMD reactions (Chan et al., 2009), the role of UPF3X/UPF3b in NMD is not clearly defined, except in bridging UPF2 to the EJC and stimulating the helicase activity of UPF1 (see the section on UPF1). In addition, its role in NMD is not absolutely required, since a downregulation of UPF3X/UPF3b is partially compensated by UPF3/UPF3a (Chan et al., 2009). Moreover, some NMD reactions, such as on the T-cell receptor mRNA, occur in the absence of UPF3/UPF3a and UPF3X/UPF3b (Chan et al., 2007), consistent with the idea that several pathways can activate NMD.

3.2.5 Suppressor of Morphogenesis in Genitalia 1 (SMG1)/ATX/ Lambda-Iota Protein Kinase C-Interacting Protein (LIP)

SMG1 is a big protein with a molecular weight of 410 kDa and a member of the phosphatidylinositol 3-kinase (PI3K)-related protein kinase (PIKK) family encoded by a gene located on the chromosome 16 (Fig. 1.17). Its role in NMD is to phosphorylate UPF1 when the ribosome is posing at a PTC and when UPF1 interacts with UPF2 and/or UPF3/UPF3X proteins. SMG1 phosphorylates the threonine 28 at the N-terminal end of UPF1 and several serines at the C-terminal end of UPF1, and in particular the serine 1096 that serves as anchor to SMG6 and SMG5-SMG7, respectively (Okada-Katsuhata et al., 2011). Since preventing the interaction between SMG1 and UPF1 impairs NMD (Hu et al., 2013), this interaction became an interesting target to identify inhibitors of NMD by looking for molecules capable of interfering with this interaction (Usuki et al., 2004).

SMG1 interacts with UPF1 independently of the presence of the other UPF proteins or the EJC. In this isolated complex, the kinase activity of SMG1 is repressed by the proteins SMG8 and SMG9 (see Section 3.2.9) (Yamashita et al., 2009). It is only when SMG1 and UPF1 are recruited to the EJC and to the translation termination complex (ribosome, eRF1, and eRF3) that SMG1 phosphorylates UPF1, once UPF2 interacts with the C-terminal part of SMG1 (Kashima et al., 2006).

Besides its activity in NMD, SMG1 as a PI3 kinase related kinase family member is also involved in other related processings, such as DNA damage response or telomere maintenance. Indeed, SMG1 is activated when cells are exposed to genotoxic stresses, leading to the phosphorylation of

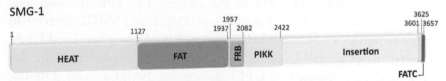

Figure 1.17 *Schematic representation of SMG1 and its functional domain.* As member of the PI3K family, SMG1 possesses a Huntington, elongation factor 3, a subunit of PP2A, and TOR1 (HEAT) repeat domain at the N-terminal part. The HEAT domain is followed by a conserved ~600 amino acids FRAP/TOR, ATM, and TRRAP (FAT) domain and a ~100 amino acids FKBP12-rapamycin-binding (FRB) domain involved in the binding with UPF2 (Melero et al., 2014). Then, the catalytic kinase domain (PIKK) of about 340 amino acids is indicated in yellow, followed by a poorly characterized insertion domain of more than 1000 amino acids (insertion). Finally, at the C-terminal end of SMG1 is the FAT C-terminal (FATC) domain required for the kinase activity of SMG1 and for protein–protein interaction (Lempiainen and Halazonetls, 2009). The amino acid position is indicated at the top.

downstream targets, such as the serine 15 of P53 (Brumbaugh et al., 2004; Gewandter et al., 2011).

In addition, SMG1 has been involved in another mechanism requiring UPF1 called Staufen-mediated mRNA decay (SMD), which is a decay pathway of specific mRNAs, recognized by the Staufen protein via a secondary structure located in the 3' UTR of these mRNAs (Park and Maquat, 2013). Indeed, downregulation of SMG1 impairs SMD and overexpression of the nonfunctional version of SMG1 also inhibits SMD (Cho et al., 2013).

Recently, two cofactors of SMG1 have been identified and called SMG8 and SMG9 (see Section 3.2.9) (Yamashita et al., 2009). Both proteins are present in the SURF complex and interact with SMG1 in order to regulate the activity of SMG1, together with some additional proteins, called RuvB-like AAA ATPase1 (RUVBL1) and RuvB-like AAA ATPase2 (RUVBL2) (Izumi et al., 2010). It is expected that the activity of SMG1 is tightly regulated, since the phosphorylation of UPF1 is the key event that will precipitate an mRNA to a fast decay.

3.2.6 SMG5/EST1B

The SMG5 gene is located on the chromosome 1, and encodes a protein of about 114 kDa (Fig. 1.18). SMG5 is found in the cytoplasm, as well as in the nucleus (Durand et al., 2007; Unterholzner and Izaurralde, 2004). In the cytoplasm, a fraction of SMG5 localizes into the processing bodies (P bodies) (Note 1.1). SMG5 harbors a nonfunctional Pilt N-terminus (PIN) domain at its C-terminal end, which is a domain found in single strand RNases (Glavan et al., 2006) and a tetratricopeptide repeat (TPR) domain at the N-terminal end of the protein. The TPR domains and, more particularly, the 14-3-3 domain of SMG5 and SMG7, interact to form a heterodimer SMG5–SMG7 that will bind phosphorylated UPF1 (Chakrabarti et al., 2014). SMG5 functions as a component of the dephosphorylation complex of UPF1 during the NMD process (Ohnishi et al., 2003).

The role of SMG5 in NMD is to interact with UPF1, together with the protein phosphatase 2A, SMG7, and SMG6, in order to promote the dephosphorylation of UPF1 (Ohnishi et al., 2003). Besides its involvement in NMD, SMG5 and SMG6 have been shown to interact with the human telomerase reverse transcriptase (hTERT) and to play a role in the maintenance of the length of telomeres (Reichenbach et al., 2003; Snow et al., 2003). This

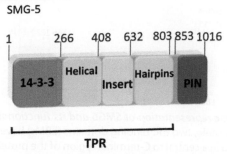

Figure 1.18 *Schematic representation of SMG5 and its functional domains.* At the N-terminal end of the protein is found the tetratricopeptide region formed by a 14-3-3 domain and a helical hairpins domain interrupted by an insertion of 220 amino acids. At the C-terminal part of the protein is the PIN domain. The amino acid position is indicated at the top.

NOTE 1.1 The Processing Bodies (P-bodies)

P-bodies have been described for the first time by following the cellular localization of the 5′–3′ exoribonuclease Xrn1 (Bashkirov et al., 1997). They were then characterized as cytoplasmic foci, containing proteins involved in RNA decay, such as Lsm proteins, the decapping enzymes, Staufen, argonaute, and proteins involved in NMD (Anderson and Kedersha, 2006; Cougot et al., 2004; Durand et al., 2007; Ingelfinger et al., 2002; Liu et al., 2005; Sheth and Parker, 2003; Unterholzner and Izaurralde, 2004; van Dijk et al., 2002). Some RNAs have been shown to be present in P-bodies such as miRNAs (Pillai et al., 2005) or PTC-containing mRNAs (Durand et al., 2007). Although ribosomal proteins are not found in P-bodies, indicating that translation does not occur in P-bodies, some proteins involved in translation have been found in these cytoplasmic foci, such as eIF4E (Anderson and Kedersha, 2006; Kedersha et al., 2005). P-bodies are dynamic structures, but cannot be considered as an organelle, since they are not limited by a membrane, and their function is still not clear, at least in mammals, unlike in yeast, where it has been demonstrated that P-bodies are the place where RNAs are degraded (Sheth and Parker, 2003). In mammals, P-bodies could be either a storage place for RNAs and decay enzymes or the site where RNA decay takes place.

---

role involves the fraction of SMG5 present in the nucleus and dedicated to a process independent of NMD, even though several NMD factors, such as UPF1, UPF2, or SMG1, take part in the maintenance of the integrity of the genome and the length of telomeres (Azzalin and Lingner, 2006a,b; Azzalin et al., 2007; Brumbaugh et al., 2004).

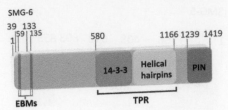

Figure 1.19 *Schematic representation of SMG6 and its functional domains.* At the N-terminal end of the protein, two EJC binding domains (EBMs) have been identified. The TPR domain is located in a central to C-terminal region of the protein and is formed by a 14-3-3 motif and a helical hairpins domain. The functional PIN domain is situated at the C-terminal end. The amino acid position is indicated at the top.

3.2.7 SMG6/EST1A/hSmg5/7a

The SMG6 gene is located on the chromosome 17 and encodes a protein of about 160 kDa. SMG6 localizes almost exclusively in the cytoplasm, where it concentrates into P-bodies (Durand et al., 2007; Unterholzner and Izaurralde, 2004). Several functional domains have been found in the SMG6 protein (Fig. 1.19). Just like for SMG5, a PIN domain is present at the C-terminal part of the protein but, unlike SMG5, this domain is functional in SMG6, since SMG6 has an RNase activity (Huntzinger et al., 2008). Indeed, SMG6 has been shown to have an endonuclease activity that is thought to cleave RNA in the vicinity of the PTC (Eberle et al., 2009; Huntzinger et al., 2008; Mascarenhas et al., 2013). In the middle of the C-terminal region of the protein, the TPR domain, and in particular a 14-3-3 like domain, serves to the interaction with phosphorylated UPF1 (Fukuhara et al., 2005). Finally, the N-terminal end of the protein interacts with the EJC via two EJC binding motifs (EBM) (Kashima et al., 2010). SMG6 also interacts with UPF1 in a phosphorylated-independent way, via the stalk and the SQ domains of UPF1 (Nicholson et al., 2014).

In the NMD process, SMG6 binds to UPF1, once UPF1 has been loaded on EJC and phosphorylated by SMG1. It is at that step that SMG6, via its endonucleolytic activity, cuts the mRNA in the vicinity of the PTC to generate two fragments with an unprotected 5′ or a 3′ end (Eberle et al., 2009; Huntzinger et al., 2008; Mascarenhas et al., 2013). These fragments are then quickly degraded by the exonucleolytic pathways, involving the exosome for the decay from the 3′ end to the 5′ end, and the exoribonucleases XRN1 or XRN2 for the decay from the 5′ to the 3′ end (Lejeune et al., 2003). Interestingly, SMG6 shows a preference in its cleavage sequence, as shown by analysis of hundreds of endogenous substrates for SMG6 cleavage (Schmidt et al., 2014b). The degenerated consensus sequence for the favored cleavage substrates of SMG6 is (U/A)(G/A)↕(C/A)N(C/U) (where N is for any ribonucleotides and ↕ symbolizes the cleavage site of SMG6) located close to the stop codon. Importantly, this study was performed on natural substrates of NMD, in which the stop codon activating NMD and its nucleotide environment have been selected during evolution to be efficiently recognized and cleaved by SMG6. Indeed, changing two nucleotides of a SMG6 cleavage site into two nucleotides that do not fit the consensus sequence strongly inhibits SMG6 cleavage. It will be essential to understand how SMG6 cleaves in the vicinity of PTCs for which the nucleotide environment likely differs from the consensus sequence.

Besides its role in NMD, SMG6 is also involved in the maintenance of the length of telomeres, as described in the SMG5 section.

3.2.8 SMG7/EST1C

The SMG7 gene is carried by the chromosome 1 and encodes a protein of about 127 kDa that localizes in both nucleus and cytoplasm, where it concentrates into P-bodies (Durand et al., 2007; Unterholzner and Izaurralde, 2004). Several domains have been identified in SMG7 (Fig. 1.20). In particular, a TPR domain is present at the N-terminal end of the protein, formed by a 14-3-3 domain responsible for the heterodimerization with SMG5 and a helical hairpins domain. The TPR domain is followed by a linker region and a proline-rich domain, called PC. The PC domain is

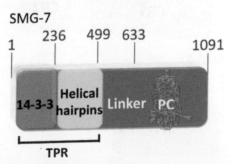

Figure 1.20 *Schematic representation of SMG7 and its functional domains.* The TPR formed by a 14-3-3 domain and a helical hairpins domain is located at the N-terminal part of the protein followed by a linker region and a C-terminal proline-rich (named PC) region. The amino acid position is indicated at the top.

responsible for the mRNA decay activity of SMG7, by stimulating decapping and deadenylation. In particular, the PC domain of SMG7 was shown to interact directly with the protein POP2, which is a catalytic subunit of the CCR4-NOT deadenylase complex (Loh et al., 2013).

SMG7, together with SMG5, interacts with phosphorylated UPF1, via its 14-3-3 like domain. This interaction occurs when UPF1 has been recruited by the EJC and has been phosphorylated by SMG1. Interestingly, SMG7 is capable to address phosphorylated UPF1 to P-bodies, indicating that a step of NMD occurs in these cytoplasmic foci (Durand et al., 2007; Fukuhara et al., 2005; Unterholzner and Izaurralde, 2004).

3.2.9 SMG8/Amplified in Breast Cancer Gene 2 and SMG9

SMG8 and SMG9 were identified after mass fingerprinting analysis of peptides coimmunoprecipitating with SMG1 (Yamashita et al., 2009). The SMG8 gene is found on chromosome 17 and encodes a protein of about 110 kDa. The SMG9 gene is carried by the chromosome 19 and encodes a protein of about 60 kDa. Some functional domains have been identified on both proteins, but these proteins have not been deeply characterized (Fig. 1.21).

SMG8 and SMG9 interact tightly with SMG1 and are thought to repress the kinase activity of SMG1. Interestingly, SMG8 interacts with SMG1 only

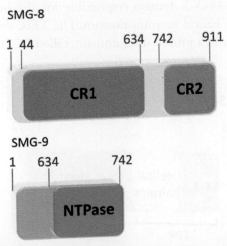

Figure 1.21 *Schematic representation of SMG8 (upper panel) and SMG9 (lower panel) and their functional domains.* SMG8 is composed by two conserved regions named CR1 and CR2. SMG9 has a putative central nucleoside triphosphatase (NTPase) domain. The amino acid position is indicated at the top.

in the presence of SMG9 and is involved in the recruitment of the SMG1 complex bound to UPF1 by the EJC and the ribosome stalled at the PTC. The regulation activity of SMG8 and SMG9 on SMG1 is completed by additional proteins, named RUVBL1 and RUVBL2 (Izumi et al., 2010).

The involvement of all these NMD factors is universally recognized and admitted, but the identification of a nonsense codon as a premature termination is still subject to debate in mammalian cells and two models are proposed (Sections 3.3 and 3.4). Each one has strong experimental arguments, but none of them can explain all the described PTC recognitions. Based on both models, a third one starts to emerge and could be closer to the truth.

3.3 EJC-Dependent Model

In the 1990s, the relative position of introns versus stop codon appears to influence the activation of NMD (Carter et al., 1996; Cheng et al., 1994; Sun and Maquat, 2000; Sun et al., 2000; Thermann et al., 1998; Zhang et al., 1998b). Indeed, at least one intron has to be present downstream of the PTC in order to elicit NMD, and with a minimal distance of 50–55 nucleotides (Zhang et al., 1998a). Interestingly, physiological stop codons are located in the last exon, except in few cases where the stop codon is carried by an other exon and at less than 50 nucleotides upstream of the last exon–exon junction for 98% of these exceptions; (Hawkins, 1988; Nagy and Maquat, 1998). It appears quickly that the intron was not the element inducing NMD by itself, but the splicing of the intron. Indeed, a protein complex has been identified at 20–24 nucleotides upstream of the majority of exon–exon junctions (Sauliere et al., 2012; Singh et al., 2012). This complex called EJC (exon junction complex) is deposited as a mark in order to signal where the splicing reaction occurred (Le Hir et al., 2000a,b). Some of its components have been detected on the RNA, as early as concomitant to the spliceosome C complex (Fig. 1.6) (Ideue et al., 2007; Reichert et al., 2002; Schmidt et al., 2014a; Singh et al., 2012; Wahl et al., 2009; Zhang and Krainer, 2007).

The EJC is a multiprotein complex of about 335 kDa composed by about 17 proteins (Table 1.4). Among these proteins are found splicing factors (SRm160, RNPS1, SAP18, Pinin, Pnn/DRS, UAP56, MLN51), proteins involved in mRNA export (UAP56, Magoh, Y14, TAP, and Aly/REF), and some core EJC proteins (Y14, MAGOH, eIF4AIII, MLN51). Due to its protein composition, EJC has been found to be involved in pre-mRNA splicing, mRNA export, translation, and NMD (Andersen et al., 2006;

Table 1.4 Composition of the EJC

	EJC's proteins	Function	References
Stable core complex	RNPS1 ⎤ SAP18 ⎬ PSAP Complex Acinus ⎟ Pinin ⎦	Pre-mRNA splicing coactivator Pre-mRNA splicing factor Transcriptional control factor Splicing coactivator	Tange et al. (2005); Murachelli et al. (2012)
	Pnn/DRS	Pre-mRNA splicing factor	Li et al. (2003)
	SRm160	Pre-mRNA splicing factor	Le Hir et al. (2001)
	UAP56	Pre-mRNA splicing and mRNA export factor	Gatfield et al. (2001); Luo et al. (2001)
	Brentsz/MLNS1	Pre-mRNA splicing factor	Degot et al. (2004); Andersen et al. (2006); Bono et al. (2004); Jackson et al. (2010)
	elF4AIII	Translation enhancer	
	Magoh	mRNA export factor	
	Y14	mRNA export factor	
	REF/Aly	mRNA export factor	Le Hir et al. (2001); Gatfield and Izaurralde (2002)
	TAP-p15	mRNA export factor	Strasser and Hurt (2000); Lejeune et al. (2002)
	UPF2	Nonsense-mediated mRNA decay factor	Le Hir et al. (2001)
	UPF3/3X	Nonsense-mediated mRNA decay factor	Kim et al. (2001); Gehring et al. (2003)
	PYM	EJC disassembly factor	Bono et al. (2004)
	SKAR	Translation preinitiation factor	Ma et al. (2008)

Bono et al., 2004; Bono and Gehring, 2011; Degot et al., 2004; Gatfield and Izaurralde, 2002; Gatfield et al., 2001; Gehring et al., 2003, 2009; Jackson et al., 2010; Kim et al., 2001; Le Hir et al., 2001; Lejeune et al., 2002; Li et al., 2003; Luo et al., 2001; Ma et al., 2008; Murachelli et al., 2012; Strasser and Hurt, 2000; Tange et al., 2005).

In NMD, the role of the EJC is thought to be the recruitment of NMD factors UPF3 or UPF3X, and then UPF2, all considered to be components of the EJC. Then, during the pioneer round or first round of translation, ribosomes remove EJCs, while they are reading the ORF (Ishigaki et al., 2001). On a wild-type mRNA, no EJCs will remain on the mRNA, when the ribosome reaches the physiological stop codon, which is the first stop codon met by the ribosome. Indeed, physiological stop codons localize either in the last exon, or at less than 50–55 nucleotides upstream of the last exon–exon junction (Hawkins, 1988; Nagy and Maquat, 1998). In the case of a PTC-containing mRNA, the first stop codon met by the ribosome is the PTC. It has to be located more than 50–55 nucleotides upstream of an exon–exon junction, that is, about 30 nucleotides upstream of an EJC, in order to be recognized as a PTC (Fig. 1.22). The minimum distance between the PTC and a downstream EJC that prevents the removal of the EJC by the ribosome seems to be about 30 nucleotides. Therefore, if the PTC is situated at less than 50–55 nucleotides upstream of the last exon–exon junction, the ribosome will remove the last EJC on the mRNA, before reaching, or when it reaches, the PTC. This PTC will not elicit NMD and the PTC-containing mRNA will be immune to NMD and will promote the synthesis of a truncated protein.

The sequential events leading to the recognition of a PTC and the decay of the mRNA carrying this PTC are as follows (Fig. 1.23). First, the EJC is deposited in the nucleus 20–24 nucleotides upstream of most of the exon–exon junctions, as a consequence of splicing (Le Hir et al., 2000a,b; Sauliere et al., 2012; Singh et al., 2012). The EJC recruits UPF3 or UPF3X in the nucleus, before the export to the cytoplasm of the mRNP (Lejeune et al., 2002). UPF2 is then recruited by UPF3 or UPF3X before the first/pioneer round of translation (Ishigaki et al., 2001). It is during the pioneer round of translation that PTCs are recognized. Indeed, the ribosome reads the mRNA and translates it into a peptide (Apcher et al., 2011). The ribosome removes all EJCs that it meets until the first stop codon. If the stop codon is in the last exon or at less than 50–55 nucleotides upstream of the last exon–exon junction, the mRNA is not subject to NMD. When the first stop codon met by the ribosome is a PTC, at least one EJC remains downstream of the PTC, since the ribosome pauses at the first stop codon. UPF1, SMG1, SMG8, and SMG9 are recruited as a complex with the release factors 1 and 3 (eRF1 and eRF3) to the ribosome stalled at the PTC, in order to form the SURF complex (Kashima et al., 2006). SMG8 and CBP80 favor the interaction between UPF1 and UPF2, in order to form the DECID complex (Hosoda et al., 2005; Yamashita et al., 2009). At

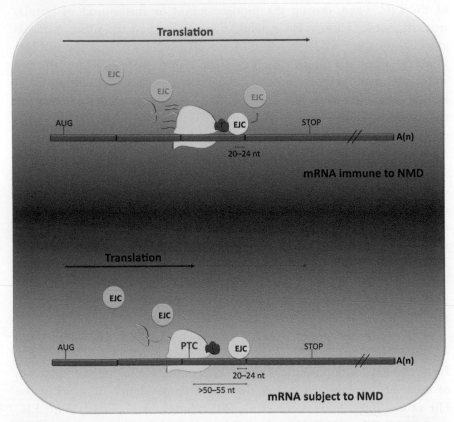

Figure 1.22 *The position of the first stop codon determines the fate of the mRNA.* For a wild-type mRNA (upper panel) during the pioneer round of translation, the ribosome translates the ORF and removes all the EJCs that it meets, until it reaches the first stop codon which is the physiological stop codon located in the last exon. No EJCs will remain on the mRNA, making this mRNA immune to NMD. In the case of a PTC containing mRNA (lower panel), the ribosome translates the ORF during the pioneer round of translation until it reaches the first stop codon, which is the PTC. The PTC is located at more than 50–55 nucleotides upstream of an exon–exon junction, making the distance between the ribosome pausing on the PTC and the downstream EJC sufficient to maintain EJC on the mRNA. This mRNA will then be subject to NMD.

that moment, UPF2 interacts with SMG1, releasing the kinase activity of SMG1 that phosphorylates UPF1. UPF2 and UPF3 or UPF3X also interact with UPF1 to free its 5′ to 3′ helicase activity. Phosphorylated UPF1 induces the leaving of the release factors 1, the release factor 3, and the ribosome. In parallel, phosphorylated UPF1 recruits the dephosphorylating complex formed by the heterodimer SMG5/SMG7 (Jonas et al., 2013; Ohnishi et al., 2003), SMG6 and the protein

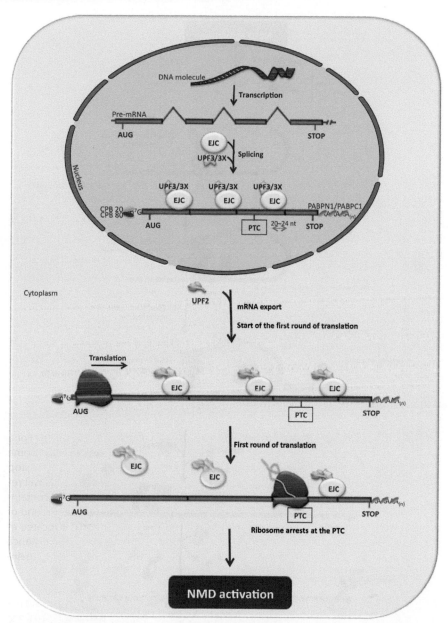

Figure 1.23 *Model of NMD activation dependent of the EJC (see text).*

(*Continued*)

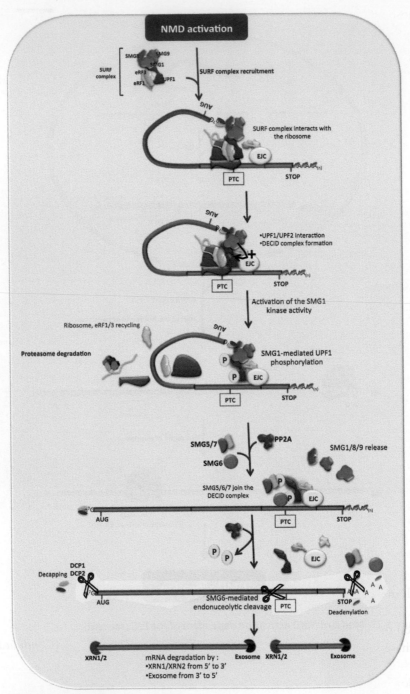

Figure 1.23 *(cont.)*

phosphatase 2A leading to the concomitant release of SMG1, SMG8, and SMG9. SMG7 is thought to be an adaptor targeting UPF1 and the PTC-containing mRNP to the P-bodies (Durand et al., 2007; Fukuhara et al., 2005; Unterholzner and Izaurralde, 2004). UPF1 is dephosphorylated by the protein phosphatase 2A and SMG6 induces an endonucleolytic cleavage in the vicinity of the PTC (Huntzinger et al., 2008; Mascarenhas et al., 2013; Schmidt et al., 2014b). UPF1, thanks to its helicase activity, removes proteins bound on the downstream fragment making it a substrate free of proteins and ready for decay by exoribonucleases, such as XRN1 or XRN2. The decapping activity and the deadenylation activity are also activated in order to release the mRNA ends from their protective proteins, and make them sensitive to the exoribonucleolytic decay (Cho et al., 2009; Lejeune et al., 2003). PTC-containing mRNAs are then degraded by different decay pathways, leading to an efficient reduction of the amount of PTC-containing mRNAs.

The involvement of EJC in NMD has been demonstrated by tethering some components of the EJC downstream of a stop codon in order to mimic the anchor of an EJC and by measuring the level of the corresponding mRNA (Gehring et al., 2008; Lykke-Andersen et al., 2000; Palacios et al., 2004). Results show that tethering Y14, MAGOH, RNPS1, or eIF4AIII downstream of the physiological stop codon of the β-globin mRNA, leads to a strong decrease of the level of the β-globin mRNA (Gehring et al., 2005). Consistent with the involvement of EJC in NMD, downregulating Y14 using siRNA impairs significantly NMD (Gehring et al., 2003).

An additional argument to support the EJC-dependent model comes from natural substrates of NMD that are wild-type genes using NMD to regulate their own expression (Mendell et al., 2004) (see Section 3.5). Some of them activate a splicing in their 3′ UTR to promote the deposit of an EJC downstream of the physiological stop codon and activate NMD. For example, the splicing factor SRSF2 (SC35) regulates its expression via NMD (Sureau et al., 2001). When the level of SRSF2 protein is abnormally high in a cell, this splicing factor activates several cryptic splicing events on the 3′ UTR of its own pre-mRNA. Those splicing events do not occur when the SRSF2 protein is not too abundant in the cell. The consequences of these splicing events are the presence of several EJCs downstream of the physiological stop codon. During the pioneer round of translation, this stop codon will then be recognized as a PTC, and will induce NMD on the SRSF2 mRNA, leading to the decrease of the level of SRSF2 protein, in order to bring it back to the physiological level. Overall, these studies demonstrate that EJC plays an essential role in NMD.

The 50–55 nucleotides rule applies to many cases of PTCs, but not to all. Examples have been described, in which EJCs are present downstream of a PTC, and should induce NMD but do not (Knezevic et al., 1995; Nagy and Maquat, 1998). In some other situations, PTCs located at less than 50–55 nucleotides upstream of the last exon–exon junction are able to elicit NMD. For instance, a nonsense mutation at 15 nucleotides upstream of the last exon–exon junction of the TCRβ gene elicits NMD, when clearly this PTC should not be able to induce it according to the 50–55 nucleotides rule (Wang et al., 2002). Several hypotheses could explain that exception, starting by the presence of an EJC downstream of the last exon–exon junction that would be deposited according to an unusual way, independently of a splicing event. Another hypothesis would be that NMD can be activated independently of the presence of EJCs (see Section 3.5). In another cases, some PTCs close to the translation initiation codon and situated at more than 50–55 nucleotides upstream of an exon–exon junction fail to elicit NMD as it has been reported for some PTCs in HNF-1beta mRNA for instance (Harries et al., 2005). The hypothesis to explain such exception to the 50–55 nucleotides rule is a possible translation reinitiation, downstream of the PTC. Translation reinitiation can be thought when PTC is close to the original translation initiation codon and if another translation initiation codon is available in the close vicinity, downstream of the PTC. Such event will result in the synthesis of a truncated protein lacking its N-terminal end.

Another argument challenges the EJC-dependent model of NMD activation, based on two recent studies showing that only a fraction of the exon–exon junctions carries an EJC at 20–24 nucleotides upstream of the splicing event (Sauliere et al., 2012; Singh et al., 2012). These studies suggest that the systematic deposit of an EJC after a spicing event is not strictly occuring and the EJC might not be a reliable mark to detect PTCs. Other studies present evidence that the signal eliciting NMD is the distance between the PABPC1 and the stop codon, rather than the presence of an EJC downstream of a PTC (see Section 3.4).

3.4 Model Involving the Distance Between the Stop Codon and the Position of the poly(A) Binding Protein C1

In yeast, *C. elegans*, or *Drosophila*, the EJC does not exist or does not play a role in NMD (Conti and Izaurralde, 2005). In these species, the size of the 3′ UTR is relatively more homogenous than in mammals. Because of this homogeneity in the size of the 3′ UTR in yeast, *C. elegans*, or *Drosophila*, it is possible that a mechanism measures it and detects when it is abnormally long to elicit NMD. In mammals, a similar model has been proposed based on the demonstration that a PTC does not activate NMD if the PABPC1 is artificially tethered to a close downstream area of the PTC (Behm-Ansmant et al., 2007; Eberle et al., 2008; Ivanov et al., 2008; Silva et al., 2008; Singh et al., 2008). The size of the 3′ UTR in mammals is much more heterogeneous than in the other species, making it difficult for the moment to understand which markers are used to measure the distance between the first stop codon of the ORF and the position of the PABPC1, and how NMD is activated in this model that does not require the EJC. The model suggests that there occurs a competition between UPF1 and the PABPC1 to interact with the release factor 3, and when the PABPC1 is close to the stop codon, UPF1 has no chance to make an interaction with eRF3 and to activate NMD (Fig. 1.24). In contrast, when the 3′ UTR is long, the distance between the stop codon and the PABPC1 favors the interaction between the release factor 3 and UPF1 rather than with PABPC1 and the activation of NMD.

However, this model has recently been challenged by the demonstration that a mutant PABPC1 unable to interact with eRF3 is still capable of inhibiting NMD when it is tethered close to a PTC. The interaction between PABPC1 and the translation initiation factor eIF4G promotes the circularization of the mRNA, which is crucial to the promotion of NMD on PTC-containing mRNA, independently of the interaction between PABPC1 and the eRF3 (Fatscher et al., 2014).

Supporting the model of the length of the 3′ UTR, the demonstration has been made that the introduction of an intron downstream of a normal stop codon is not sufficient to elicit NMD (Singh et al., 2008). Another supporting argument comes from the discovery that UPF1 concentrates in the 3′ UTR of mRNAs without the requirement of an EJC (Hogg and Goff, 2010; Kurosaki et al., 2014; Kurosaki and Maquat, 2013; Zund et al., 2013). UPF1 and in particular phosphorylated UPF1 would be the marker to differentiate between a normal and an abnormally long 3′ UTR according to the number of phosphorylated UPF1 present downstream of the stop codon. Knowing that tethering UPF1 downstream of a normal termination codon elicits NMD without recruitment of UPF2, UPF3/UPF3X, and the EJC (Lykke-Andersen et al., 2000), this suggests that the EJC might not be absolutely necessary for NMD, as long as UPF1 finds a way to anchor downstream of the stop codon. Consistent with the non requirement of the presence of EJCs to elicit NMD, two studies showed

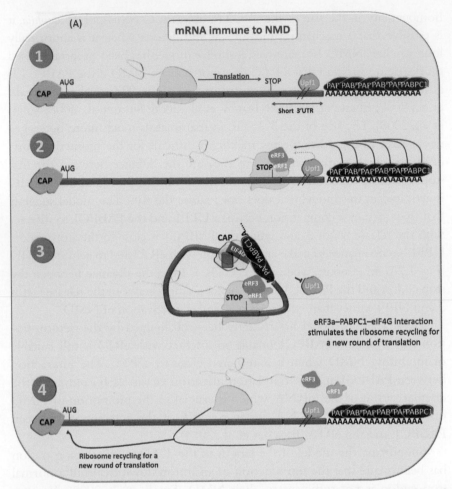

Figure 1.24 *Model for NMD activation involving the distance between the PABPC1 and the stop codon.* For mRNAs immune to NMD (A), the distance between the first stop codon and the PABPC1 is not recognized as long. The competition between UPF1 and PABPC1 for the interaction with eRF3 is in favor of the PABPC1 proteins. PABPC1 interacts with eIF4G and the eRF3 stimulates the recycling of the ribosome for new rounds of translation. For mRNAs subject to NMD

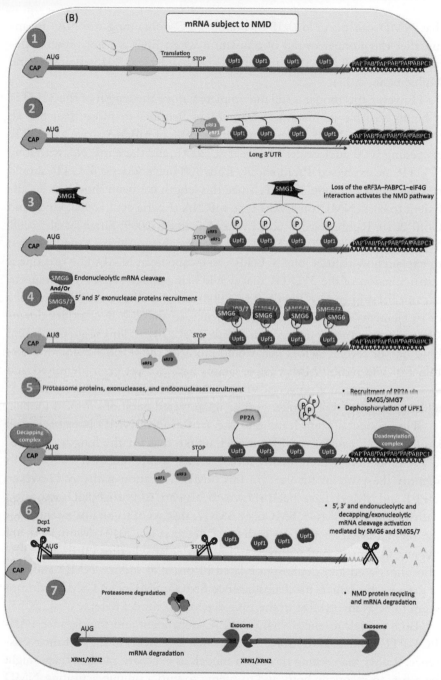

Figure 1.24 (cont.) (B), the distance between the first stop codon and the PABPC1 is recognized as long. eRF3 has more chance then to interact with UPF1 than with PABPC1. UPF1 is then phosphorylated by SMG1 and SMG5/SMG7 and/or SMG6 are recruited to dephosphorylate UPF1, via the protein phosphatase 2A (PP2A), inducing the decay of the mRNA by exo- and/or endonucleolytic cleavage.

that NMD can occur on eIF4E-bound PTC-containing mRNAs mean-ing after the pioneer round of translation when EJCs are not present any-more on the mRNA (Durand and Lykke-Andersen, 2013; Rufener and Mühlemann, 2013).

However, this model is still uncompleted, since the length of the 3′ UTR is highly variable in mammals, from 21 nucleotides to more than 8.5 kb (Mignone et al., 2002). Indeed, more than 30% of mRNAs have a 3′ UTR exceeding 1000 nucleotides (Pesole et al., 2000), and these mRNA with long 3′ UTR are expressed. Paradoxically, from 800 nucleotides, a 3′ UTR should be considered as a long 3′ UTR, since this length has been shown to be suf-ficient to elicit NMD on a wild-type mRNA (Eberle et al., 2008; Hogg and Goff, 2010; Rebbapragada and Lykke-Andersen, 2009; Singh et al., 2008; Yepiskoposyan et al., 2011). Until now, two mechanisms have been proposed to explain how natural long 3′ UTR can escape from NMD. The first one is a structuration of the 3′ UTR that would bring closer PABPC1 proteins to the normal stop codon (Eberle et al., 2008) and the other one is the presence of *cis* elements, such as A/U rich sequence, in the 200 nucleotides down-stream of the normal stop codon (Toma et al., 2015). This sequence would bind one or several factors that would promote the normal translation ter-mination and inhibit NMD. These factors have not yet been identified, and this mechanism is not universal, since an A/U rich sequence is not present in all long natural 3′ UTR, raising some unanswered questions for this model.

The sequential steps leading to the activation of NMD have been well studied in the case of the EJC model, unlike that for the length of the 3′ UTR model, for which further investigations will be necessary in order to identify the proteins involved in this NMD activation pathway. However, UPF1 and SMG1 have been shown to be part of both NMD activation pathways, unlike SMG5, SMG6, or SMG7, that seem to be not essential for the second NMD activation pathway (Metze et al., 2013). Many steps are missing for this second pathway, and in particular whether UPF1 has to be phosphorylated and dephosphorylated in order to induce NMD and, if it is the case, by which mechanism, since SMG5, SMG6, and SMG7 do not seem to be as involved as in the EJC-dependent model (Metze et al., 2013).

Both models to elicit NMD, the EJC-dependent and the distance PAB-PC1-PTC models, have strong experimental arguments to support their exis-tence. Rather than seeing those two models as exclusive models, there might be a way to reconcile both, and to merge toward a unique activating NMD model that would explain the wide situations of mRNAs eliciting NMD. The idea would be that both models could coexist, and the EJC would be an activator/potentiator of NMD, but would not be absolutely required (Buhler et al., 2006; Metze et al., 2013). Indeed, as for the other organisms, the length of the 3' UTR would be the determinant to elicit NMD in mammalian cells. During evolution, the EJC has been integrated in the NMD process likely to optimize the recognition of PTCs. For instance, except for the TCRβ gene (Wang et al., 2002), NMD efficiency is generally not dependent of the location of the PTC in the ORF in mammals (Cheng et al., 1994; Zhang et al., 1998b), unlike in yeast, where a polar effect is clearly observed with a high NMD efficiency for the PTCs situated at the most distal position from the 3' end of the mRNA (Leeds et al., 1991; Losson and Lacroute, 1979; Pelsy and Lacroute, 1984). Since mRNAs can be extremely long in mammals, the recognition of PTCs located at the beginning of the ORF in particular need some beacons likely to bring the NMD machinery at a reasonable distance from any putative PTCs. Then, the recognition of the PTC is achieved via EJCs, if there are some present downstream of the PTC, according to the EJC model described previously, or according to the model involving the distance between the PABPC1 and the stop codon, when no EJCs are present downstream of the PTC. The recognition of PTCs via the length of the distance between the PTC and the PABPC1 could represent an ancestral mechanism conserved during evolution, still active in mammals, and optimized by the PTC recognition EJC-dependent. The resulting NMD activation model would be that PTCs are recognized first during the pioneer round of translation according to the EJC-dependent model and the fraction of PTC-containing mRNAs that would escape from this first recognition would be subject to a second analysis via the distance PABPC1-PTC model.

Although the merge model can explain most of NMD activations, some specific cases still resist and cannot be explained by that model. For example, PTCs in the last exon often are immune to NMD like for β globin gene in which PTCs in the last exon lead to a dominant negative form of β-thalassemia, by producing truncated nonfunctional β globin chains (Thein et al., 1990). According to the EJC dependent model, PTCs in the last exon are not subject to NMD and generate truncated proteins which is consistent with what is observed in this situation. It is more difficult to understand under the merge model law since the absence of EJC downstream to the PTC should be compensate by the detection of an increased length of the 3' UTR, compared to the WT mRNA which is not what is observed. Another particular example of PTCs that do not fit the activation of NMD by any existing models comes from PTCs in the COL10A1 gene (Fang et al., 2013; Tan et al., 2008). This gene harbors three exons, with the last exon that represents the majority of the ORF. PTCs in this exon elicit NMD, when located at the 3′ end of the exon and do not activate NMD when located in the rest of the exon 3 (Fig. 1.25). According to the EJC dependent model, none of these PTCs should activate NMD and the most distant PTCs from the 3′ UTR should be the ones that elicit NMD, according to the model involving the distance between the first stop codon and the PABPC1 or the merge model. To explain that situation, an additional model has been suggested, in which ribosomes translating the ORF, from the translation initiation codon until the PTC, protect that region of the

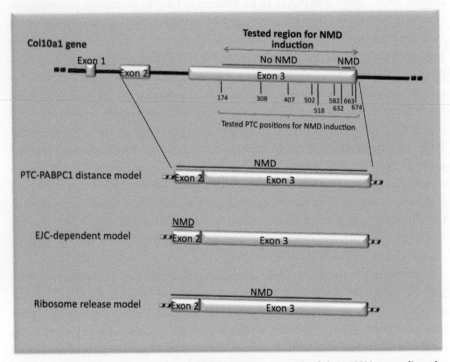

Figure 1.25 *NMD activation in the last exon (the exon 3) of the mRNA, encoding the Col10a1 protein.* On this mRNA, PTCs in the last exon elicit NMD when they are at the 3′ end of the exon 3, but not when they are located in the rest of the exon (upper part). For the PTC-PABPC1 distance model (or the merge model), if PTCs at the end of the exon elicit NMD, upstream PTCs should also activate NMD, making this model irrelevant for this mRNA. For the EJC-dependent model, only PTCs located at more than 50–55 nucleotides upstream of the exon 3 should elicit NMD, which is not the case here, making this model not suitable for this mRNA. Another proposed model is about the ribosome release model but, according to that model, upstream PTCs should also activate NMD (see text), also making this model not capable to explain fully the NMD activation on this model.

mRNA from being degraded by RNases, unlike the mRNA region from the PTC to the 3′ end that is not protected by ribosomes and, therefore, accessible to RNases (Brogna and Wen, 2009). Even that model does not fit the experimental data 100% because not all PTCs in the exon 3 induce NMD. It is always possible to hypothesize about a secondary structure that would bring the PABPC1 close in the case of the upper PTC, and not for the downstream PTC because, for these latter, ribosome would read most of this exon and would destabilize the secondary structure, allowing NMD to be activated. Such exceptions need to be further studied, in order to understand why they do not fit the majority of the cases, and then allow improving the existing merge model.

3.5 Natural Substrates of NMD

The role of NMD in eliminating mRNAs harboring a PTC due to a frameshift or nonsense mutation is essential for the cell, in order to prevent the synthesis of failure proteins. Another role of NMD was more recently highlighted and focuses on the regulation of some gene expression. This role was first identified after downregulating hUPF1 or hUPF2 in the human HeLa cells, using siRNAs (Mendell et al., 2004). The transcriptomic analysis leads to the conclusion that about 5% of the human genome is upregulated when NMD is inactivated. It is possible to organize most of these genes into categories according to their mode of regulation involving NMD (Fig. 1.26). Indeed, genes with an upstream open reading frame (uORF), harboring an intron in the 3′ UTR, encoding a selenoprotein, subject to alternative splicing leading to the introduction of a PTC by intron retention or frameshift, and genes targeted by transposon elements that introduce a PTC or induce a frameshift, are natural substrates of NMD. According to external and/or internal parameters, these genes are expressed by escaping from NMD or they are repressed if they are subject to NMD.

For example, in the case of mRNA harboring an uORF, the mRNA escapes from NMD, as long as only the major open reading frame (mORF) is translated, and the uORF is ignored by the translation machinery. When the uORF is translated, the stop codon of the uORF is recognized as a PTC and will activate NMD. The consequence is a decrease of the synthesis of the protein translated from the mORF. It is, therefore, an efficient regulation pathway related to the translation of the uORF.

For mRNAs encoding a selenoprotein, a UGA codon is present in the ORF, and can be recognized by a transfer RNA carrying a serine transformed into a selenocysteine, in the presence of selenium. Such recognition

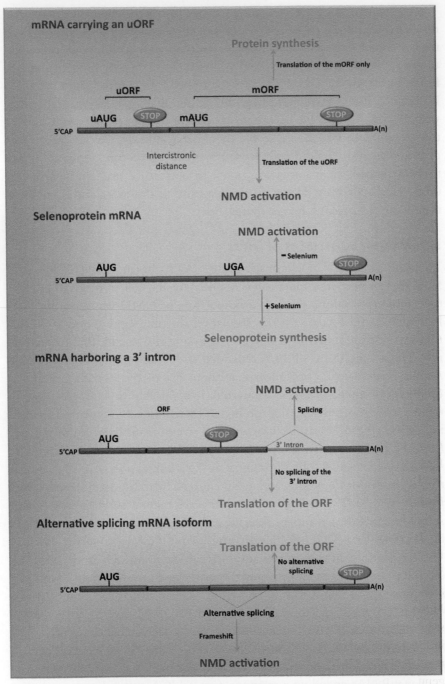

Figure 1.26 *Categories of natural substrates of NMD.* Exons are represented by a rectangle and introns by a thick horizontal line. uAUG, uORF, mAUG, and mORF stand for the upstream translation initiation codon, upstream open reading frame, major translation initiation codon, and major open reading frame. STOP shows the position of physiological stop codons.

of a stop codon UGA to incorporate a selenocysteine is extremely controlled, and requires the presence of a *cis* element on the mRNA, called SeC insertion sequence (SECIS), and the presence of factors such as the SECIS binding protein 2 (SBP2) and the translation factor eEFsec. When selenium is present in the cellular environment, such UGA codon leads to the incorporation of a selenocysteine. In contrast, when the cellular environment is poor in selenium, the UGA codon will be recognized as a PTC and will activate NMD in order to prevent the synthesis of selenoproteins.

A major source of production of PTC is by alternative splicing. About 35% of mRNA isoforms generated by alternative splicing harbor a PTC (Green et al., 2003). The use of alternative splicing to incorporate a PTC and to promote the silencing of a gene is called regulated unproductive splicing and translation mechanism (RUST) (Lewis et al., 2003). This mechanism has been conserved during the evolution from yeast to human, highlighting its requirement for the homeostasis of gene expression (Lareau et al., 2007a). Many genes encoding splicing factors belong to the genes using RUST, and in particular the serine/arginine rich protein (SR protein) family (Lareau et al., 2007b; Lewis et al., 2003; Sureau et al., 2001; Wollerton et al., 2004). The members of this family regulate their expression by affecting the splicing events occurring on their own pre-mRNA, and in particular by incorporating a poison cassette exon containing a PTC (Fig. 1.27) (Lareau and Brenner, 2015; Lareau et al., 2007b; Lejeune et al., 2001; Sureau et al., 2001).

Genes encoding NMD factors are also using NMD as a gene regulation pathway (Yepiskoposyan et al., 2011). However, genes encoding NMD factors used either an upstream of ORF or a long 3' UTR to activate NMD, suggesting a complex way to regulate the activation of NMD on these mRNAs, involving translation and likely secondary structure of the 3' UTR.

Regulation of the gene expression by NMD is universally admitted but the number of genes using NMD as a regulation pathway is likely less than 5% of the human genome. Indeed, 5% of the human genome is upregulated after downregulation of UPF1 or UPF2 by siRNAs (Mendell et al., 2004). However, about 5% of the human genome is also downregulated under the same experimental conditions, indicating that approach measured direct and indirect effect of an inhibition of NMD. Therefore, in the 5% of genes upregulated after NMD inhibition, a proportion is upregulated because another gene product was affected by the NMD inhibition. Consistent with that, another study analyzed some of the genes that were found to be upregulated at the origin, when NMD is inhibited; it was also found that upregulation

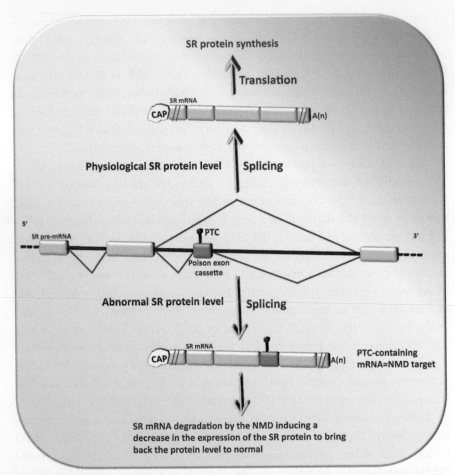

Figure 1.27 *Regulation of SR protein expression via alternative splicing and NMD.* When the level of the SR protein is normal, the mRNA does not include the poison exon cassette containing the PTC. This mRNA will be translated to generate the SR protein. When the level of the SR protein is abnormally high, the poison exon cassette is retained in the mRNA, introducing a PTC in the ORF. That mRNA will be degraded by NMD leading to an absence of the SR protein synthesis and a decrease in the level of that protein.

already occurs at the pre-mRNA level, suggesting that is a transcription effect and not an effect via NMD (Viegas et al., 2007). Indeed, it is difficult to distinguish genes directly regulated by NMD and genes whose expression will be affected because of the absence of NMD, due to the overexpression of a direct NMD substrate (Fig. 1.28). The amount of natural targets of NMD is therefore expected to be less than 5% of the human genome.

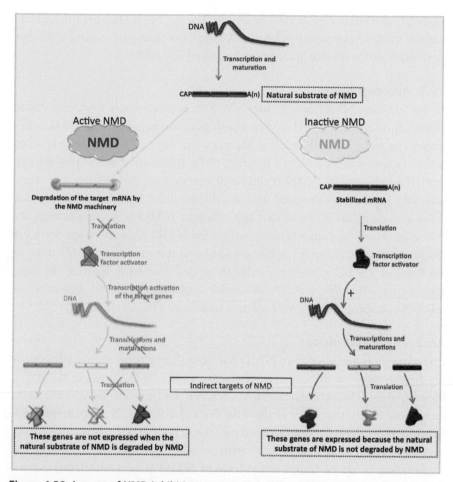

Figure 1.28 *Impact of NMD inhibition on gene expression.* When NMD is active, natural substrates of NMD are degraded by NMD. The product of these natural substrates of NMD can regulate the expression of other genes, like transcription factors. Those genes are not expressed when NMD degrade the mRNA encoding the transcription factor (left side), unlike when NMD is inhibited (right side). A global transcriptional analysis under NMD inhibition will select natural substrates of NMD and genes with an expression dependent of the level of natural substrates of NMD, even though these last ones are not subject to NMD.

3.6 Regulation

NMD, like any other quality control mechanism, has to be tightly controlled in order to fill its functions accurately. Though the molecular details of the mechanism have been extensively studied, the regulation of NMD begins to be investigated. This regulation can affect the entire NMD mechanism, and often reflects a cell status (cell differentiation or cell death, for instance) or modulate the efficiency of one or several specific NMD factors (phosphorylation, miRNA, or autoregulation). We are now going to explore different regulation pathways that have been described for NMD.

3.6.1 Autoregulation

As we just saw, NMD factors use NMD to regulate their own expressing (Yepiskoposyan et al., 2011). Such autoregulation suggests that NMD efficiency has to be limited and, if it becomes too high or too low, the level of NMD factors will be adjusted in order to be maintained at the physiologic level. If a low level of NMD could be deleterious for the cell because some mRNAs would be expressed and would generate harmful proteins, it is more surprising to imagine that too efficient NMD is not a situation allowed in cells. That suggests some targets for NMD have to escape for a yet unknown reason. However, it has already been reported that NMD reduces the level of a PTC-containing mRNA to 5–25% of the level of the corresponding wild-type mRNA (Kuzmiak and Maquat, 2006), but most of mRNAs escaping from NMD are not translated (You et al., 2007).

3.6.2 Tissue Specificity

Interestingly, the efficiency of NMD has been shown to be regulated according to the tissue or the cell type (Bateman et al., 2003; Viegas et al., 2007). Indeed, some PTC-containing mRNA levels are found to be variable from one tissue to another, as it is the case for the collagen X in patients with Schmid metaphyseal chondrodysplasia, for which NMD was efficient in cartilage cells and absent in noncartilage cells (Bateman et al., 2003). An explanation could be found at the level of NMD factors or related factors, since a comparison of them in different cell types showed that the amount of RNPS1, for example, correlates with the efficiency of NMD, suggesting that a measure of RNPS1 could be a marker of the efficiency of NMD in a cell type or in a tissue (Viegas et al., 2007).

3.6.3 Inhibition During Apoptosis

Under specific cell conditions, NMD can also be regulated. Very recently, it has been demonstrated that NMD is inhibited during apoptosis due to the cleavage by caspases of the NMD factors UPF1 and UPF2, at least, but not UPF3X (Jia et al., 2015; Popp and Maquat, 2015). The specific cleavage of UPF1 and UPF2 suggests that the action of caspases against NMD during apoptosis is targeted, rather than dictated by a nonselective cleavage of components of a pathway. To support this idea, the N-terminal fragments of UPF1 or UPF2 generated by caspases have a slight but reproducible apoptotic effect (Jia et al., 2015). Interestingly, the NMD inhibition occurring during apoptosis leads to the expression of genes regulated by NMD, and proapoptotic for some of them (Popp and Maquat, 2015). These studies suggest that NMD has to be inhibited during apoptosis, and plays an active role in cell death progression.

3.6.4 miRNA

MicroRNAs (miRNAs) regulate a wide range of genes, either by affecting the stability of the mRNA or by inhibiting translation on the mRNA that it targets (Bartel, 2009; Jonas and Izaurralde, 2015). Each miRNA targets a set of genes, and several miRNAs can influence the expression of the same gene. Among these miRNAs, miR128 regulates the expression of UPF1 (Bruno et al., 2011).

The regulation of NMD by the miRNA pathway is likely not only restricted to the regulation of UPF1 expression, since the protein AGO2, a catalytic subunit of the RISC (RNA-induced silencing complex), preferentially binds mRNAs carrying EJCs and CBP80/CBP20, that is, before the pioneer round of translation, so before PTC can elicit NMD. This binding blocks translation on the bound mRNA and prevents its decay by NMD (Choe et al., 2010, 2011). It is therefore expected that some NMD substrates escape from NMD; thanks to miRNAs recognition and binding (Note 1.2).

NOTE 1.2 Repression Mechanism of Gene Expression by miRNAs

MicroRNAs (miRNAs) are small noncoding RNAs of about 22 nucleotides long. They repress gene expression by inhibiting translation and/or promoting mRNA decay. For that, the miRNA generally binds to the 3′ UTR of its target mRNA and recruits an argonaute protein and a GW182 protein in order to form the micro RNA-induced silencing complex (miRISC). The miRISC inhibits translation by interfering with the assembly or the function of the translation initiation complex eIF4F (composed by eIF4G, eIF4E, and eIF4A). The GW182 protein recruits the pan2/pan3 deadenylase complex and the CCR4-Not 3′–5′ exoribonuclease complex, and also the decapping complex and the 5′–3′ exoribonuclease XRN1, in order to degrade the miRNA target mRNA. About 1500 miRNAs are present in human cells and each miRNA targets hundred mRNAs, suggesting that a significant proportion of human genes are regulated by miRNAs. MiRNAs are also regulated in order to silence the expression of specific genes at the accurate moment and for a specific period of time. Indeed, the translation on some mRNAs can be repressed by miRNAs and then be subject to a bulk of translation.

3.6.5 Phosphorylation

Posttranslational modifications play a crucial role in NMD and in particular phosphorylation. UPF1 and UPF2 are phosphorylated and, at least for UPF1, the phosphorylation and dephosphorylation steps are required for the activation of NMD. The phosphorylation of UPF1 occurs on several serine/threonine-glutamine motifs (at least T28, S1078, S1096, S1116) thanks to the kinase SMG1 (Matsuoka et al., 2007; Ohnishi et al., 2003; Okada-Katsuhata et al., 2011; Yamashita et al., 2001). The dephosphorylation of UPF1 occurs in a protein complex composed by the proteins SMG5, SMG7, SMG6, and the protein phosphatase 2A (PP2A).

UPF2 is also a phosphoprotein (Chiu et al., 2003), suggesting that the functions of that protein could be modulated via the posttranslational process of phosphorylation and dephosphorylation. However, this regulation has not yet been studied in mammals, making the role of UPF2 phosphorylation in NMD hypothetic.

3.6.6 Regulation by Availability of NMD Factors

Many if not all proteins involved in NMD play multiroles in cells. If a pathway becomes very active and requires the essential amount of factors also involved in NMD, making the amount of these factors limiting for other pathways, such situation is expected to have a consequence on the efficiency of NMD. The best example to illustrate this is the protein UPF1 that plays a role in different pathways, and in particular in NMD and Staufen-mediated mRNA decay (SMD) (Kim et al., 2005). UPF1 interacts with both UPF2 (an NMD factor) and Staufen (an SMD factor), and both proteins share the same binding site on UPF1 (Gong et al., 2009). Indeed, if the level of UPF2 decreases in the cell, SMD will become more efficient and, in contrast, NMD will be less efficient. To exemplify the relative link between NMD and SMD efficiency, that competition has been reported to occur during myogenesis, for instance. During such differentiation, UPF1 is found to interact more with Staufen than with UPF2, favoring SMD and impacting NMD (Gong et al., 2009).

3.7 UPF2, UPF3X/UPF3b Independent Pathway

Although many proteins have been identified as essential factors of NMD, such as UPF and SMG proteins, some NMD reactions do not always required all these proteins. In particular, UPF2 has been shown to be dispensable, according to the composition of the EJC (Gehring et al., 2005). Indeed, if the EJC is lacking in RNPS1, NMD can occur even in the absence of

UPF2. In contrast, the presence of RNPS1 in the EJC requires UPF2 to activate NMD. Although the composition of EJCs seems to be homogenous on endogenous mRNAs (Singh et al., 2012), a possibility of heterogeneous EJC composition cannot be definitively excluded and would support the idea that NMD can be activated by different pathways, involving UPF2 or not.

UPF3X/UPF3b has also been shown to not be absolutely required for NMD since downregulating this factor does not impair NMD (Chan et al., 2009). This can be explained by the fact that UPF3/UPF3a, the product of the paralog gene of UPF3X/UPF3b, can functionally replace UPF3X/UPF3b. However, as it is going to be explained in the next chapter, the replacement of UPF3X/UPF3b by UPF3/UPF3a is only partial, and a lack of UPF3X/UPF3b leads to severe retard mental disorders, suggesting that the replacement might not be efficient in all types of cells, tissues, or NMD reactions.

The fact that some UPF proteins can be dispensable in some NMD reactions demonstrates that NMD can be activated by different pathways. The role, the specificity, the composition, and the overlapping functions between these different pathways remain widely unknown.

3.8 Pathologies Associated with NMD Defaults

In mammals, the abolition of NMD is lethal as demonstrated in mice by a knockout of UPF1 gene that leads to an embryonic lethality at 3.5 day post coitum (dpc) (Medghalchi et al., 2001). UPF1 is not the only NMD factor to be shown to induce embryonic lethality when it is missing. For instance, mouse embryos lacking in SMG1 expression die around 8.5 dpc (McIlwain et al., 2010). Although this lethality cannot be exclusively allocated to NMD impairment since both UPF1 and SMG1 proteins have been shown to participate in different cellular pathways (see chapter: General Aspects Related to Nonsense Mutations; Sections 3.2.1 and 3.2.5). It has been shown that NMD plays a crucial role in the development of tissues and organs. For example, the expression of a dominant negative version of UPF1 (UPF1 R843C) impairs NMD and promotes a developmental arrest of fetal thymocytes (Frischmeyer–Guerrerio et al., 2011). Thymocytes are expected to be affected by a lack of NMD since the locus encoding the T-cell receptor is subject to a programmed DNA rearrangement and two-thirds of these rearrangements lead to the introduction of a PTC (Mallick et al., 1993). As another example, in a conditional UPF2 knockout mouse, UPF2 has been demonstrated to be required for terminal liver differentiation during the development and for liver regeneration in adult liver (Thoren et al., 2010).

It is therefore expected that human pathologies associated with a default in NMD factors are rare.

Since some pathologies associated with a lack of the function of one NMD factor have been described, cells sometimes find a way to compensate the missing function. Indeed, UPF3X/UPF3b can be dispensable for the embryogenesis survival, likely due to the presence of the UPF3/UPF3a protein that can functionally replace UPF3X/UPF3b (Chan et al., 2009). However, the replacement of UPF3X/UPF3b by UPF3/UPF3a is not total, and causes the development of pathologies. For example, a point mutation transforming the Lysine 367 into an asparagine in UPF3X/UPF3b gene leads to an X-linked intellectual disability (Nguyen et al., 2012; Tzschach et al., 2015) and other neurological disorders, such as schizophrenia, autism, or attention-deficit hyperactivity (Addington et al., 2011; Laumonnier et al., 2010; Tarpey et al., 2007). To understand the mechanism behind the development of neuronal-related pathologies with a default in the UPF3X/UPF3b functions, the expression of UPF3X/UPF3b has been studied. Indeed, UPF3X/UPF3b is expressed during the brain development at about 10 folds higher than UPF3/UPF3a (Jolly et al., 2013). In the same study, the authors also show that under the loss of UPF3X/UPF3b, the neural progenitor cells divide more and differentiate less than in the presence of UPF3X/UPF3b, suggesting a role of this protein in the cell differentiation program. A transcriptomic analysis also revealed that in the absence of UPF3X/UPF3b, 16 genes highly expressed in neuron cells are deregulated (SIX3, TMOD2, NRCAM, or ROBO1 for instance), indicating the impact in some gene expression of the absence of UPF3X/UPF3b.

Sometimes the mutation at the origin of the pathology does not affect a gene involved in NMD directly, but is located in a gene that regulates the expression of an NMD factor. It is, for example, found in some cases of craniofacial dysmorphisms in which a nonsense mutation was found in the SATB2 gene, leading to the generation of a truncated SATB2 protein with a dominant negative effect, since this protein dimerizes to form a functional transcriptional complex. One of the targets of SATB2 is the UPF3X/UPF3b gene, explaining why the patients with that mutation have the same phenotype as patients with mutations in the UPF3X/UPF3b gene (Leoyklang et al., 2013), and explaining why the truncated protein is synthesized from a PTC-containing mRNA, since NMD is inhibited.

UPF3X/UPF3b is not the only gene involved in NMD that can be found as the origin of a human pathology and in particular in intellectual disorders. A genome analysis of patients with intellectual disorders reveals that, besides mutations in UPF3X/UPF3b, genomic alterations can be found in UPF2, UPF3/UPF3A, SMG6, eIF4AIII, or RNPS1 (Nguyen et al., 2013). These genes can be likely found in patients because their inactivation does not tremendously affects cell viability. It is also likely because NMD can be activated by different combinations of proteins, as we saw, before making possible the impairment of one of them, without affecting significantly the expression of genes using NMD to regulate their expression.

4 CORRECTION OF NONSENSE MUTATIONS, A CASE OF TARGETED THERAPY

Before the development of personalized or targeted medicine, all patients with the same pathology received generally the same treatment. By deciphering the molecular defaults at the origin of genetic diseases, it became necessary to design treatments according to the molecular event inducing the pathology, in order to improve safety and efficiency. The genetic background is also a key component of personalized medicine, since it has already been shown to influence the efficacy of a treatment, explaining why the same drug promotes variable effects on a cohort of patients (Cazzola et al., 2015; De Mattia et al., 2015; Graziani and Nistico, 2015). Molecular pathways such as NMD have been demonstrated to be modulated from one patient to another, providing also an explanation at least on the influence of the genetic background for genetic diseases related with nonsense mutations (Haas et al., 2015; Linde et al., 2007; Nguyen et al., 2013; Viegas et al., 2007; Welch et al., 2007).

Nonsense mutations are adapted to targeted therapies because such mutations can be found in any ORFs, and the molecular consequences are shared for all mutant genes, unlike a missense, a deletion, or insertion mutations, for example, that will be dependent of the gene and the mutant protein. Indeed, the molecular consequences of a nonsense mutation are the silencing of the mutant gene, due to the specific and fast decay of the corresponding mRNA by NMD. Therefore, the development of strategies to correct a nonsense mutation in a specific gene may apply to nonsense mutations in various genes, since such therapeutic strategy will be developed independently of the function of the mutant gene. However, as it will be described later, some parameters related to the position of the mutation influence the correction efficiency and in particular when readthrough strategy is involved.

Correction of nonsense mutations aims to rescue the expression of genes harboring a nonsense mutation that will apply to a fraction of patients with a genetic disease. Although, based on the in vitro and ex vivo data, molecules capable of rescuing the expression of genes harboring a nonsense mutation show variability in their efficiency according to the cell type or the nucleotide environment around the mutation, it is expected that one molecule will be able to treat many patients. For all these reasons, therapies focusing on the correction of nonsense mutations are targeted therapies, rather than a development of personalized medicine.

REFERENCES

Addington, A.M., Gauthier, J., Piton, A., Hamdan, F.F., Raymond, A., Gogtay, N., Miller, R., Tossell, J., Bakalar, J., Inoff-Germain, G., et al., 2011. A novel frameshift mutation in UPF3B identified in brothers affected with childhood onset schizophrenia and autism spectrum disorders. Mol. Psychiatry 16, 238–239.

Anczukow, O., Ware, M.D., Buisson, M., Zetoune, A.B., Stoppa-Lyonnet, D., Sinilnikova, O.M., Mazoyer, S., 2008. Does the nonsense-mediated mRNA decay mechanism prevent the synthesis of truncated BRCA1, CHK2, and p 53 proteins? Hum. Mutat. 29, 65–73.

Andersen, C.B., Ballut, L., Johansen, J.S., Chamieh, H., Nielsen, K.H., Oliveira, C.L., Pedersen, J.S., Seraphin, B., Le Hir, H., Andersen, G.R., 2006. Structure of the exon junction core complex with a trapped DEAD-box ATPase bound to RNA. Science 313, 1968–1972.

Anderson, P., Kedersha, N., 2006. RNA granules. J. Cell Biol. 172, 803–808.

Apcher, S., Daskalogianni, C., Lejeune, F., Manoury, B., Imhoos, G., Heslop, L., Fahraeus, R., 2011. Major source of antigenic peptides for the MHC class I pathway is produced during the pioneer round of mRNA translation. Proc. Natl. Acad. Sci. USA 108, 11572–11577.

Araki, K., Nagata, K., 2011. Protein folding and quality control in the ER. Cold Spring Harb. Perspect. Biol. 3, a007526.

Ashiya, M., Grabowski, P.J., 1997. A neuron-specific splicing switch mediated by an array of pre-mRNA repressor sites: evidence of a regulatory role for the polypyrimidine tract binding protein and a brain-specific PTB counterpart. RNA 3, 996–1015.

Atkinson, J., Martin, R., 1994. Mutations to nonsense codons in human genetic disease: implications for gene therapy by nonsense suppressor tRNAs. Nucleic Acids Res. 22, 1327–1334.

Azzalin, C.M., Lingner, J., 2006a. The double life of UPF1 in RNA and DNA stability pathways. Cell Cycle 5, 1496–1498.

Azzalin, C.M., Lingner, J., 2006b. The human RNA surveillance factor UPF1 is required for S phase progression and genome stability. Curr. Biol. 16, 433–439.

Azzalin, C.M., Reichenbach, P., Khoriauli, L., Giulotto, E., Lingner, J., 2007. Telomeric repeat containing RNA and RNA surveillance factors at mammalian chromosome ends. Science 318, 798–801.

Bartel, D.P., 2009. MicroRNAs: target recognition and regulatory functions. Cell 136, 215–233.

Bashkirov, V.I., Scherthan, H., Solinger, J.A., Buerstedde, J.M., Heyer, W.D., 1997. A mouse cytoplasmic exoribonuclease (mXRN1p) with preference for G4 tetraplex substrates. J. Cell Biol. 136, 761–773.

Bateman, J.F., Freddi, S., Nattrass, G., Savarirayan, R., 2003. Tissue-specific RNA surveillance? Nonsense-mediated mRNA decay causes collagen X haploinsufficiency in Schmid metaphyseal chondrodysplasia cartilage. Hum. Mol. Genet. 12, 217–225.

Bechara, E.G., Sebestyen, E., Bernardis, I., Eyras, E., Valcarcel, J., 2013. RBM5, 6, and 10 differentially regulate NUMB alternative splicing to control cancer cell proliferation. Mol. Cell 52, 720–733.

Behm-Ansmant, I., Gatfield, D., Rehwinkel, J., Hilgers, V., Izaurralde, E., 2007. A conserved role for cytoplasmic poly(A)-binding protein 1 (PABPC1) in nonsense-mediated mRNA decay. EMBO J. 26, 1591–1601.

Bhuvanagiri, M., Schlitter, A.M., Hentze, M.W., Kulozik, A.E., 2010. NMD: RNA biology meets human genetic medicine. Biochem. J. 430, 365–377.

Black, D.L., 1992. Activation of c-src neuron-specific splicing by an unusual RNA element in vivo and in vitro. Cell 69, 795–807.

Blencowe, B.J., 2006. Alternative splicing: new insights from global analyses. Cell 126, 37–47.

Boehm, V., Haberman, N., Ottens, F., Ule, J., Gehring, N.H., 2014. 3′ UTR length and messenger ribonucleoprotein composition determine endocleavage efficiencies at termination codons. Cell Rep. 9, 555–568.

Bono, F., Gehring, N.H., 2011. Assembly, disassembly and recycling: the dynamics of exon junction complexes. RNA Biol. 8, 24–29.

Bono, F., Ebert, J., Unterholzner, L., Guttler, T., Izaurralde, E., Conti, E., 2004. Molecular insights into the interaction of PYM with the Mago-Y14 core of the exon junction complex. EMBO Rep. 5, 304–310.

Boon, K.L., Grainger, R.J., Ehsani, P., Barrass, J.D., Auchynnikava, T., Inglehearn, C.F., Beggs, J.D., 2007. prp8 mutations that cause human retinitis pigmentosa lead to a U5 snRNP maturation defect in yeast. Nat. Struct. Mol. Biol. 14, 1077–1083.

Bourgeois, C.F., Lejeune, F., Stevenin, J., 2004. Broad specificity of SR (serine/arginine) proteins in the regulation of alternative splicing of pre-messenger RNA. Prog. Nucleic. Acid Res. Mol. Biol. 78, 37–88.

Brogna, S., Wen, J., 2009. Nonsense-mediated mRNA decay (NMD) mechanisms. Nat. Struct. Mol. Biol. 16, 107–113.

Brumbaugh, K.M., Otterness, D.M., Geisen, C., Oliveira, V., Brognard, J., Li, X., Lejeune, F., Tibbetts, R.S., Maquat, L.E., Abraham, R.T., 2004. The mRNA surveillance protein hSMG-1 functions in genotoxic stress response pathways in mammalian cells. Mol. Cell 14, 585–598.

Bruno, I.G., Karam, R., Huang, L., Bhardwaj, A., Lou, C.H., Shum, E.Y., Song, H.W., Corbett, M.A., Gifford, W.D., Gecz, J., et al., 2011. Identification of a microRNA that activates gene expression by repressing nonsense-mediated RNA decay. Mol. Cell 42, 500–510.

Buchwald, G., Ebert, J., Basquin, C., Sauliere, J., Jayachandran, U., Bono, F., Le Hir, H., Conti, E., 2010. Insights into the recruitment of the NMD machinery from the crystal structure of a core EJC-UPF3b complex. Proc. Natl. Acad. Sci. USA 107, 10050–10055.

Buhler, M., Steiner, S., Mohn, F., Paillusson, A., Muhlemann, O., 2006. EJC-independent degradation of nonsense immunoglobulin-mu mRNA depends on 3′ UTR length. Nat. Struct. Mol. Biol. 13, 462–464.

Busch, A., Hertel, K.J., 2013. HEXEvent: a database of Human EXon splicing Events. Nucleic Acids Res. 41, D118–D124.

Caceres, J.F., Stamm, S., Helfman, D.M., Krainer, A.R., 1994. Regulation of alternative splicing in vivo by overexpression of antagonistic splicing factors. Science 265, 1706–1709.

Cali, B.M., Kuchma, S.L., Latham, J., Anderson, P., 1999. smg-7 is required for mRNA surveillance in Caenorhabditis elegans. Genetics 151, 605–616.

Carastro, L.M., Tan, C.K., Selg, M., Jack, H.M., So, A.G., Downey, K.M., 2002. Identification of delta helicase as the bovine homolog of HUPF1: demonstration of an interaction with the third subunit of DNA polymerase delta. Nucleic Acids Res. 30, 2232–2243.

Carter, M.S., Li, S., Wilkinson, M.F., 1996. A splicing-dependent regulatory mechanism that detects translation signals. EMBO J. 15, 5965–5975.

Cazzola, M., Rogliani, P., Sanduzzi, A., Matera, M.G., 2015. Influence of ethnicity on response to asthma drugs. Expert Opin. Drug Metab. Toxicol. 11, 1089–1097.

Chakrabarti, S., Jayachandran, U., Bonneau, F., Fiorini, F., Basquin, C., Domcke, S., Le Hir, H., Conti, E., 2011. Molecular mechanisms for the RNA-dependent ATPase activity of Upf1 and its regulation by Upf2. Mol. Cell 41, 693–703.

Chakrabarti, S., Bonneau, F., Schussler, S., Eppinger, E., Conti, E., 2014. Phospho-dependent and phospho-independent interactions of the helicase UPF1 with the NMD factors SMG5-SMG7 and SMG6. Nucleic Acids Res. 42, 9447–9460.

Chamieh, H., Ballut, L., Bonneau, F., Le Hir, H., 2008. NMD factors UPF2 and UPF3 bridge UPF1 to the exon junction complex and stimulate its RNA helicase activity. Nat. Struct. Mol. Biol. 15, 85–93.

Chan, R.C., Black, D.L., 1995. Conserved intron elements repress splicing of a neuron-specific c-src exon in vitro. Mol. Cell. Biol. 15, 6377–6385.

Chan, W.K., Huang, L., Gudikote, J.P., Chang, Y.F., Imam, J.S., MacLean, II, J.A., Wilkinson, M.F., 2007. An alternative branch of the nonsense-mediated decay pathway. EMBO J. 26, 1820–1830.

Chan, W.K., Bhalla, A.D., Le Hir, H., Nguyen, L.S., Huang, L., Gecz, J., Wilkinson, M.F., 2009. A UPF3-mediated regulatory switch that maintains RNA surveillance. Nat. Struct. Mol. Biol. 16, 747–753.

Chang, Y.F., Imam, J.S., Wilkinson, M.F., 2007. The nonsense-mediated decay RNA surveillance pathway. Annu. Rev. Biochem. 76, 51–74.

Chawla, R., Redon, S., Raftopoulou, C., Wischnewski, H., Gagos, S., Azzalin, C.M., 2011. Human UPF1 interacts with TPP1 and telomerase and sustains telomere leading-strand replication. EMBO J. 30, 4047–4058.

Cheng, J., Belgrader, P., Zhou, X., Maquat, L.E., 1994. Introns are *cis* effectors of the nonsense-codon-mediated reduction in nuclear mRNA abundance. Mol. Cell. Biol. 14, 6317–6325.

Chiou, N.T., Lynch, K.W., 2014. Mechanisms of spliceosomal assembly. Methods Mol. Biol. 1126, 35–43.

Chiu, S.Y., Serin, G., Ohara, O., Maquat, L.E., 2003. Characterization of human Smg5/7a: a protein with similarities to *Caenorhabditis elegans* SMG5 and SMG7 that functions in the dephosphorylation of Upf1. RNA 9, 77–87.

Chiu, S.Y., Lejeune, F., Ranganathan, A.C., Maquat, L.E., 2004. The pioneer translation initiation complex is functionally distinct from but structurally overlaps with the steady-state translation initiation complex. Genes Dev. 18, 745–754.

Cho, H., Kim, K.M., Kim, Y.K., 2009. Human proline-rich nuclear receptor coregulatory protein 2 mediates an interaction between mRNA surveillance machinery and decapping complex. Mol. Cell 33, 75–86.

Cho, H., Han, S., Park, O.H., Kim, Y.K., 2013. SMG1 regulates adipogenesis via targeting of staufen1-mediated mRNA decay. Biochim. Biophys. Acta 1829, 1276–1287.

Choe, J., Cho, H., Lee, H.C., Kim, Y.K., 2010. microRNA/Argonaute 2 regulates nonsense-mediated messenger RNA decay. EMBO Rep. 11, 380–386.

Choe, J., Cho, H., Chi, S.G., Kim, Y.K., 2011. Ago2/miRISC-mediated inhibition of CBP80/20-dependent translation and thereby abrogation of nonsense-mediated mRNA decay require the cap-associating activity of Ago2. FEBS Lett. 585, 2682–2687.

Choe, J., Ahn, S.H., Kim, Y.K., 2014. The mRNP remodeling mediated by UPF1 promotes rapid degradation of replication-dependent histone mRNA. Nucleic Acids Res. 42, 9334–9349.

Chou, M.Y., Rooke, N., Turck, C.W., Black, D.L., 1999. hnRNP H is a component of a splicing enhancer complex that activates a c-src alternative exon in neuronal cells. Mol. Cell. Biol. 19, 69–77.

Clerici, M., Deniaud, A., Boehm, V., Gehring, N.H., Schaffitzel, C., Cusack, S., 2013. Structural and functional analysis of the three MIF4G domains of nonsense-mediated decay factor UPF2. Nucleic Acids Res. 42, 2673–2686.

Coelho, M.B., Smith, C.W., 2014. Regulation of alternative pre-mRNA splicing. Methods Mol. Biol. 1126, 55–82.

Cohen-Eliav, M., Golan-Gerstl, R., Siegfried, Z., Andersen, C.L., Thorsen, K., Orntoft, T.F., Mu, D., Karni, R., 2013. The splicing factor SRSF6 is amplified and is an oncoprotein in lung and colon cancers. J. Pathol. 229, 630–639.

Conti, E., Izaurralde, E., 2005. Nonsense-mediated mRNA decay: molecular insights and mechanistic variations across species. Curr. Opin. Cell Biol. 17, 316–325.

Cougot, N., Babajko, S., Seraphin, B., 2004. Cytoplasmic foci are sites of mRNA decay in human cells. J. Cell Biol. 165, 31–40.

Culbertson, M.R., Underbrink, K.M., Fink, G.R., 1980. Frameshift suppression *Saccharomyces cerevisiae*. II. Genetic properties of group II suppressors. Genetics 95, 833–853.

De Mattia, E., Cecchin, E., Toffoli, G., 2015. Pharmacogenomics of intrinsic and acquired pharmacoresistance in colorectal cancer: toward targeted personalized therapy. Drug Resist. Updat. 20, 39–62.

Degot, S., Le Hir, H., Alpy, F., Kedinger, V., Stoll, I., Wendling, C., Seraphin, B., Rio, M.C., Tomasetto, C., 2004. Association of the breast cancer protein MLN51 with the exon junction complex via its speckle localizer and RNA binding module. J. Biol. Chem. 279, 33702–33715.

Delpy, L., Sirac, C., Magnoux, E., Duchez, S., Cogne, M., 2004. RNA surveillance downregulates expression of nonfunctional kappa alleles and detects premature termination within the last kappa exon. Proc. Natl. Acad. Sci. USA 101, 7375–7380.

Dorard, C., de Thonel, A., Collura, A., Marisa, L., Svrcek, M., Lagrange, A., Jego, G., Wanherdrick, K., Joly, A.L., Buhard, O., Gobbo, J., Penard-Lacronique, V., Zouali, H., Tubacher, E., Kirzin, S., Selves, J., Milano, G., Etienne-Grimaldi, M.C., Bengrine-Lefèvre, L., Louvet, C., Tournigand, C., Lefèvre, J.H., Parc, Y., Tiret, E., Fléjou, J.F., Gaub, M.P., Garrido, C., Duval, A., 2011. Expression of a mutant HSP110 sensitizes colorectal cancer cells to chemotherapy and improves disease prognosis. Nat. Med. 17, 1283–1289.

Dreyfuss, G., Matunis, M.J., Pinol-Roma, S., Burd, C.G., 1993. hnRNP proteins and the biogenesis of mRNA. Annu. Rev. Biochem. 62, 289–321.

Durand, S., Cougot, N., Mahuteau-Betzer, F., Nguyen, C.H., Grierson, D.S., Bertrand, E., Tazi, J., Lejeune, F., 2007. Inhibition of nonsense-mediated mRNA decay (NMD) by a new chemical molecule reveals the dynamic of NMD factors in P-bodies. J. Cell Biol. 178, 1145–1160.

Durand, S., Lykke-Andersen, J., 2013. Nonsense-mediated mRNA decay occurs during eIF4F-dependent translation in human cells. Nat. Struct. Mol. Biol. 20, 702–709.

Eberle, A.B., Stalder, L., Mathys, H., Orozco, R.Z., Muhlemann, O., 2008. Posttranscriptional gene regulation by spatial rearrangement of the 3' untranslated region. PLoS Biol. 6, e92.

Eberle, A.B., Lykke-Andersen, S., Muhlemann, O., Jensen, T.H., 2009. SMG6 promotes endonucleolytic cleavage of nonsense mRNA in human cells. Nat. Struct. Mol. Biol. 16, 49–55.

Fang, Y., Bateman, J.F., Mercer, J.F., Lamande, S.R., 2013. Nonsense-mediated mRNA decay of collagen-emerging complexity in RNA surveillance mechanisms. J. Cell Sci. 126, 2551–2560.

Fatscher, T., Boehm, V., Weiche, B., Gehring, N.H., 2014. The interaction of cytoplasmic poly(A)-binding protein with eukaryotic initiation factor 4G suppresses nonsense-mediated mRNA decay. RNA 20, 1579–1592.

Fiorini, F., Boudvillain, M., Le Hir, H., 2013. Tight intramolecular regulation of the human Upf1 helicase by its N- and C-terminal domains. Nucleic Acids Res. 41, 2404–2415.

Fourati, Z., Roy, B., Millan, C., Coureux, P.D., Kervestin, S., van Tilbeurgh, H., He, F., Uson, I., Jacobson, A., Graille, M., 2014. A highly conserved region essential for NMD in the Upf2 N-terminal domain. J. Mol. Biol. 426, 3689–3702.

Franks, T.M., Singh, G., Lykke-Andersen, J., 2010. Upf1 ATPase-dependent mRNP disassembly is required for completion of nonsense-mediated mRNA decay. Cell 143, 938–950.

Frischmeyer-Guerrerio, P.A., Montgomery, R.A., Warren, D.S., Cooke, S.K., Lutz, J., Sonnenday, C.J., Guerrerio, A.L., Dietz, H.C., 2011. Perturbation of thymocyte development in nonsense-mediated decay (NMD)-deficient mice. Proc. Natl. Acad. Sci. USA 108, 10638–10643.

Fu, X.D., 1995. The superfamily of arginine/serine-rich splicing factors. RNA 1, 663–680.

Fu, X.D., Ares, Jr., M., 2014. Context-dependent control of alternative splicing by RNA-binding proteins. Nat. Rev. Genet. 15, 689–701.

Fu, Y., Huang, B., Shi, Z., Han, J., Wang, Y., Huangfu, J., Wu, W., 2013. SRSF1 and SRSF9 RNA binding proteins promote Wnt signalling-mediated tumorigenesis by enhancing beta-catenin biosynthesis. EMBO Mol. Med. 5, 737–750.

Fukuhara, N., Ebert, J., Unterholzner, L., Lindner, D., Izaurralde, E., Conti, E., 2005. SMG7 is a 14-3-3-like adaptor in the nonsense-mediated mRNA decay pathway. Mol. Cell 17, 537–547.

Galej, W.P., Nguyen, T.H., Newman, A.J., Nagai, K., 2014. Structural studies of the spliceosome: zooming into the heart of the machine. Curr. Opin. Struct. Biol. 25, 57–66.

Galiana-Arnoux, D., Lejeune, F., Gesnel, M.C., Stevenin, J., Breathnach, R., Del Gatto-Konczak, F., 2003. The CD44 alternative v9 exon contains a splicing enhancer responsive to the SR proteins 9G8, ASF/SF2, and SRp20. J. Biol. Chem. 278, 32943–32953.

Gatfield, D., Izaurralde, E., 2002. REF1/Aly and the additional exon junction complex proteins are dispensable for nuclear mRNA export. J. Cell Biol. 159, 579–588.

Gatfield, D., Le Hir, H., Schmitt, C., Braun, I.C., Kocher, T., Wilm, M., Izaurralde, E., 2001. The DExH/D box protein HEL/UAP56 is essential for mRNA nuclear export in *Drosophila*. Curr. Biol. 11, 1716–1721.

Gehring, N.H., Neu-Yilik, G., Schell, T., Hentze, M.W., Kulozik, A.E., 2003. Y14 and hUpf3b form an NMD-activating complex. Mol. Cell 11, 939–949.

Gehring, N.H., Kunz, J.B., Neu-Yilik, G., Breit, S., Viegas, M.H., Hentze, M.W., Kulozik, A.E., 2005. Exon-junction complex components specify distinct routes of nonsense-mediated mRNA decay with differential cofactor requirements. Mol. Cell 20, 65–75.

Gehring, N.H., Hentze, M.W., Kulozik, A.E., 2008. Tethering assays to investigate nonsense-mediated mRNA decay activating proteins. Methods Enzymol. 448, 467–482.

Gehring, N.H., Lamprinaki, S., Kulozik, A.E., Hentze, M.W., 2009. Disassembly of exon junction complexes by PYM. Cell 137, 536–548.

Gewandter, J.S., Bambara, R.A., O'Reilly, M.A., 2011. The RNA surveillance protein SMG1 activates p53 in response to DNA double-strand breaks but not exogenously oxidized mRNA. Cell Cycle 10, 2561–2567.

Glavan, F., Behm-Ansmant, I., Izaurralde, E., Conti, E., 2006. Structures of the PIN domains of SMG6 and SMG5 reveal a nuclease within the mRNA surveillance complex. EMBO J. 25, 5117–5125.

Golan-Gerstl, R., Cohen, M., Shilo, A., Suh, S.S., Bakacs, A., Coppola, L., Karni, R., 2011. Splicing factor hnRNP A2/B1 regulates tumor suppressor gene splicing and is an oncogenic driver in glioblastoma. Cancer Res. 71, 4464–4472.

Golling, G., Amsterdam, A., Sun, Z., Antonelli, M., Maldonado, E., Chen, W., Burgess, S., Haldi, M., Artzt, K., Farrington, S., et al., 2002. Insertional mutagenesis in zebrafish rapidly identifies genes essential for early vertebrate development. Nat. Genet. 31, 135–140.

Goncalves, V., Matos, P., Jordan, P., 2008. The beta-catenin/TCF4 pathway modifies alternative splicing through modulation of SRp20 expression. RNA 14, 2538–2549.

Gong, C., Kim,Y.K.,Woeller, C.F.,Tang,Y., Maquat, L.E., 2009. SMD and NMD are competitive pathways that contribute to myogenesis: effects on PAX3 and myogenin mRNAs. Genes Dev. 23, 54–66.

Gonzalez-Hilarion, S., Beghyn,T., Jia, J., Debreuck, N., Berte, G., Mamchaoui, K., Mouly,V., Gruenert, D.C., Deprez, B., Lejeune, F., 2012. Rescue of nonsense mutations by amlexanox in human cells. Orphanet J. Rare Dis. 7, 58.

Graziani, M., Nistico, R., 2015. Gender differences in pharmacokinetics and pharmacodynamics of methadone substitution therapy. Front. Pharmacol. 6, 122.

Green, R.E., Lewis, B.P., Hillman, R.T., Blanchette, M., Lareau, L.F., Garnett, A.T., Rio, D.C., Brenner, S.E., 2003. Widespread predicted nonsense-mediated mRNA decay of alternatively-spliced transcripts of human normal and disease genes. Bioinformatics 19 (Suppl. 1), i118–i121.

Haas, M., Vlcek, V., Balabanov, P., Salmonson, T., Bakchine, S., Markey, G., Weise, M., Schlosser-Weber, G., Brohmann, H.,Yerro, C.P., et al.,2015. European Medicines Agency review of ataluren for the treatment of ambulant patients aged 5 years and older with Duchenne muscular dystrophy resulting from a nonsense mutation in the dystrophin gene. Neuromuscul. Disord. 25, 5–13.

Han, K.,Yeo, G., An, P., Burge, C.B., Grabowski, P.J., 2005. A combinatorial code for splicing silencing: UAGG and GGGG motifs. PLoS Biol. 3, e158.

Harries, L.W., Bingham, C., Bellanne-Chantelot, C., Hattersley, A.T., Ellard, S., 2005. The position of premature termination codons in the hepatocyte nuclear factor -1 beta gene determines susceptibility to nonsense-mediated decay. Hum. Genet. 118, 214–224.

Hawkins, J.D., 1988. A survey on intron and exon lengths. Nucleic Acids Res. 16, 9893–9908.

He, X., Ee, P.L., Coon, J.S., Beck,W.T., 2004. Alternative splicing of the multidrug resistance protein 1/ATP binding cassette transporter subfamily gene in ovarian cancer creates functional splice variants and is associated with increased expression of the splicing factors PTB and SRp20. Clin. Cancer Res. 10, 4652–4660.

He, X., Arslan,A.D., Pool, M.D., Ho,T.T., Darcy, K.M., Coon, J.S., Beck,W.T., 2011. Knockdown of splicing factor SRp20 causes apoptosis in ovarian cancer cells and its expression is associated with malignancy of epithelial ovarian cancer. Oncogene 30, 356–365.

Hodgkin, J., Papp,A., Pulak, R.,Ambros,V.,Anderson, P., 1989. A new kind of informational suppression in the nematode *Caenorhabditis elegans*. Genetics 123, 301–313.

Hogg, J.R., Goff, S.P., 2010. Upf1 senses 3′ UTR length to potentiate mRNA decay. Cell 143, 379–389.

Hosoda, N., Kim,Y.K., Lejeune, F., Maquat, L.E., 2005. CBP80 promotes interaction of Upf1 with Upf2 during nonsense-mediated mRNA decay in mammalian cells. Nat. Struct. Mol. Biol. 12, 893–901.

Hosoda, N., Lejeune, F., Maquat, L.E., 2006. Evidence that poly(A) binding protein C1 binds nuclear pre-mRNA poly(A) tails. Mol. Cell. Biol. 26, 3085–3097.

House,A.E., Lynch, K.W., 2008. Regulation of alternative splicing: more than just the ABCs. J. Biol. Chem. 283, 1217–1221.

Hu, J., Li,Y., Li, P., 2013. MARVELD1 inhibits nonsense-mediated RNA decay by repressing serine phosphorylation of UPF1. PLoS One 8, e68291.

Hui, J., Stangl, K., Lane,W.S., Bindereif,A., 2003. HnRNP L stimulates splicing of the eNOS gene by binding to variable-length CA repeats. Nat. Struct. Biol. 10, 33–37.

Hung, L.H., Heiner, M., Hui, J., Schreiner, S., Benes,V., Bindereif,A., 2008. Diverse roles of hnRNP L in mammalian mRNA processing: a combined microarray and RNAi analysis. RNA 14, 284–296.

Huntzinger, E., Kashima, I., Fauser, M., Sauliere, J., Izaurralde, E., 2008. SMG6 is the catalytic endonuclease that cleaves mRNAs containing nonsense codons in metazoan. RNA 14, 2609–2617.

Hwang, J., Sato, H., Tang, Y., Matsuda, D., Maquat, L.E., 2010. UPF1 association with the cap-binding protein, CBP80, promotes nonsense-mediated mRNA decay at two distinct steps. Mol. Cell 39, 396–409.

Ideue, T., Sasaki, Y.T., Hagiwara, M., Hirose, T., 2007. Introns play an essential role in splicing-dependent formation of the exon junction complex. Genes Dev. 21, 1993–1998.

Imamachi, N., Tani, H., Akimitsu, N., 2012. Up-frameshift protein 1 (UPF1): multitalented entertainer in RNA decay. Drug Discov. Ther. 6, 55–61.

Ingelfinger, D., Arndt-Jovin, D.J., Luhrmann, R., Achsel, T., 2002. The human LSm1-7 proteins colocalize with the mRNA-degrading enzymes Dcp1/2 and Xrnl in distinct cytoplasmic foci. RNA 8, 1489–1501.

Ishigaki, Y., Li, X., Serin, G., Maquat, L.E., 2001. Evidence for a pioneer round of mRNA translation: mRNAs subject to nonsense-mediated decay in mammalian cells are bound by CBP80 and CBP20. Cell 106, 607–617.

Isken, O., Maquat, L.E., 2007. Quality control of eukaryotic mRNA: safeguarding cells from abnormal mRNA function. Genes Dev. 21, 1833–1856.

Ivanov, P.V., Gehring, N.H., Kunz, J.B., Hentze, M.W., Kulozik, A.E., 2008. Interactions between UPF1, eRFs, PABP and the exon junction complex suggest an integrated model for mammalian NMD pathways. EMBO J. 27, 736–747.

Izumi, N., Yamashita, A., Iwamatsu, A., Kurata, R., Nakamura, H., Saari, B., Hirano, H., Anderson, P., Ohno, S., 2010. AAA+ proteins RUVBL1 and RUVBL2 coordinate PIKK activity and function in nonsense-mediated mRNA decay. Sci. Signal 3, ra27.

Jackson, R.J., Hellen, C.U., Pestova, T.V., 2010. The mechanism of eukaryotic translation initiation and principles of its regulation. Nat. Rev. Mol. Cell Biol. 11, 113–127.

Jia, R., Li, C., McCoy, J.P., Deng, C.X., Zheng, Z.M., 2010. SRp20 is a proto-oncogene critical for cell proliferation and tumor induction and maintenance. Int. J. Biol. Sci. 6, 806–826.

Jia, J., Furlan, A., Gonzalez-Hilarion, S., Leroy, C., Gruenert, D.C., Tulasne, D., Lejeune, F., 2015. Caspases shutdown nonsense-mediated mRNA decay during apoptosis. Cell Death Differ. 22, 1754–1763.

Johnson, J.M., Castle, J., Garrett-Engele, P., Kan, Z., Loerch, P.M., Armour, C.D., Santos, R., Schadt, E.E., Stoughton, R., Shoemaker, D.D., 2003. Genome-wide survey of human alternative pre-mRNA splicing with exon junction microarrays. Science 302, 2141–2144.

Jolly, L.A., Homan, C.C., Jacob, R., Barry, S., Gecz, J., 2013. The UPF3B gene, implicated in intellectual disability, autism, ADHD and childhood onset schizophrenia regulates neural progenitor cell behaviour and neuronal outgrowth. Hum. Mol. Genet. 22, 4673–4687.

Jonas, S., Izaurralde, E., 2015. Towards a molecular understanding of microRNA-mediated gene silencing. Nat. Rev. Genet. 16, 421–433.

Jonas, S., Weichenrieder, O., Izaurralde, E., 2013. An unusual arrangement of two 14-3-3-like domains in the SMG5-SMG7 heterodimer is required for efficient nonsense-mediated mRNA decay. Genes Dev. 27, 211–225.

Kadlec, J., Izaurralde, E., Cusack, S., 2004. The structural basis for the interaction between nonsense-mediated mRNA decay factors UPF2 and UPF3. Nat. Struct. Mol. Biol. 11, 330–337.

Kalsotra, A., Cooper, T.A., 2011. Functional consequences of developmentally regulated alternative splicing. Nat. Rev. Genet. 12, 715–729.

Karam, R., Wengrod, J., Gardner, L.B., Wilkinson, M.F., 2013. Regulation of nonsense-mediated mRNA decay: implications for physiology and disease. Biochim. Biophys. Acta 1829, 624–633.

Karni, R., de Stanchina, E., Lowe, S.W., Sinha, R., Mu, D., Krainer, A.R., 2007. The gene encoding the splicing factor SF2/ASF is a proto-oncogene. Nat. Struct. Mol. Biol. 14, 185–193.

Kashima, I., Yamashita, A., Izumi, N., Kataoka, N., Morishita, R., Hoshino, S., Ohno, M., Dreyfuss, G., Ohno, S., 2006. Binding of a novel SMG-1-Upf1-eRF1-eRF3 complex

(SURF) to the exon junction complex triggers Upf1 phosphorylation and nonsense-mediated mRNA decay. Genes Dev. 20, 355–367.

Kashima, I., Jonas, S., Jayachandran, U., Buchwald, G., Conti, E., Lupas, A.N., Izaurralde, E., 2010. SMG6 interacts with the exon junction complex via two conserved EJC-binding motifs (EBMs) required for nonsense-mediated mRNA decay. Genes Dev. 24, 2440–2450.

Kedersha, N., Stoecklin, G., Ayodele, M., Yacono, P., Lykke-Andersen, J., Fritzler, M.J., Scheuner, D., Kaufman, R.J., Golan, D.E., Anderson, P., 2005. Stress granules and processing bodies are dynamically linked sites of mRNP remodeling. J. Cell Biol. 169, 871–884.

Kervestin, S., Jacobson, A., 2012. NMD: a multifaceted response to premature translational termination. Nat. Rev. Mol. Cell Biol. 13, 700–712.

Khodor, Y.L., Menet, J.S., Tolan, M., Rosbash, M., 2012. Cotranscriptional splicing efficiency differs dramatically between Drosophila and mouse. RNA 18, 2174–2186.

Kilchert, C., Vasiljeva, L., 2013. mRNA quality control goes transcriptional. Biochem. Soc. Trans. 41, 1666–1672.

Kim, V.N., Kataoka, N., Dreyfuss, G., 2001. Role of the nonsense-mediated decay factor hUpf3 in the splicing-dependent exon–exon junction complex. Science 293, 1832–1836.

Kim, Y.K., Furic, L., Desgroseillers, L., Maquat, L.E., 2005. Mammalian Staufen1 recruits Upf1 to specific mRNA 3′ UTRs so as to elicit mRNA decay. Cell 120, 195–208.

Knezevic, V., Ranson, M., Mackem, S., 1995. The organizer-associated chick homeobox gene, Gnot1, is expressed before gastrulation and regulated synergistically by activin and retinoic acid. Dev. Biol. 171, 458–470.

Kuchino, Y., Muramatsu, T., 1996. Nonsense suppression in mammalian cells. Biochimie 78, 1007–1015.

Kurosaki, T., Maquat, L.E., 2013. Rules that govern UPF1 binding to mRNA 3′ UTRs. Proc. Natl. Acad. Sci. USA 110, 3357–3362.

Kurosaki, T., Li, W., Hoque, M., Popp, M.W., Ermolenko, D.N., Tian, B., Maquat, L.E., 2014. A post-translational regulatory switch on UPF1 controls targeted mRNA degradation. Genes Dev. 28, 1900–1916.

Kuzmiak, H.A., Maquat, L.E., 2006. Applying nonsense-mediated mRNA decay research to the clinic: progress and challenges. Trends Mol. Med. 12, 306–316.

Lander, E.S., Linton, L.M., Birren, B., Nusbaum, C., Zody, M.C., Baldwin, J., Devon, K., Dewar, K., Doyle, M., FitzHugh, W., et al., 2001. Initial sequencing and analysis of the human genome. Nature 409, 860–921.

Lareau, L.F., Brenner, S.E., 2015. Regulation of splicing factors by alternative splicing and NMD is conserved between kingdoms yet evolutionarily flexible. Mol. Biol. Evol. 32, 1072–1079.

Lareau, L.F., Brooks, A.N., Soergel, D.A., Meng, Q., Brenner, S.E., 2007a. The coupling of alternative splicing and nonsense-mediated mRNA decay. Adv. Exp. Med. Biol. 623, 190–211.

Lareau, L.F., Inada, M., Green, R.E., Wengrod, J.C., Brenner, S.E., 2007b. Unproductive splicing of SR genes associated with highly conserved and ultraconserved DNA elements. Nature 446, 926–929.

Laumonnier, F., Shoubridge, C., Antar, C., Nguyen, L.S., Van Esch, H., Kleefstra, T., Briault, S., Fryns, J.P., Hamel, B., Chelly, J., et al., 2010. Mutations of the UPF3B gene, which encodes a protein widely expressed in neurons, are associated with nonspecific mental retardation with or without autism. Mol. Psychiatry 15, 767–776.

Le Hir, H., Seraphin, B., 2008. EJCs at the heart of translational control. Cell 133, 213–216.

Le Hir, H., Izaurralde, E., Maquat, L.E., Moore, M.J., 2000a. The spliceosome deposits multiple proteins 20–24 nucleotides upstream of mRNA exon–exon junctions. EMBO J. 19, 6860–6869.

Le Hir, H., Moore, M.J., Maquat, L.E., 2000b. Pre-mRNA splicing alters mRNP composition: evidence for stable association of proteins at exon–exon junctions. Genes Dev. 14, 1098–1108.

Le Hir, H., Gatfield, D., Izaurralde, E., Moore, M.J., 2001. The exon–exon junction complex provides a binding platform for factors involved in mRNA export and nonsense-mediated mRNA decay. EMBO J. 20, 4987–4997.

Lee, Y., Rio, D.C., 2015. Mechanisms and regulation of alternative pre-mRNA splicing. Annu. Rev. Biochem. 84, 291–323.

Leeds, P., Peltz, S.W., Jacobson, A., Culbertson, M.R., 1991. The product of the yeast UPF1 gene is required for rapid turnover of mRNAs containing a premature translational termination codon. Genes Dev. 5, 2303–2314.

Leeds, P., Wood, J.M., Lee, B.S., Culbertson, M.R., 1992. Gene products that promote mRNA turnover in *Saccharomyces cerevisiae*. Mol. Cell. Biol. 12, 2165–2177.

Lefave, C.V., Squatrito, M., Vorlova, S., Rocco, G.L., Brennan, C.W., Holland, E.C., Pan, Y.X., Cartegni, L., 2011. Splicing factor hnRNPH drives an oncogenic splicing switch in gliomas. EMBO J. 30, 4084–4097.

Lejeune, F., Maquat, L.E., 2005. Mechanistic links between nonsense-mediated mRNA decay and pre-mRNA splicing in mammalian cells. Curr. Opin. Cell Biol. 17, 309–315.

Lejeune, F., Cavaloc, Y., Stevenin, J., 2001. Alternative splicing of intron 3 of the serine/arginine-rich protein 9G8 gene. Identification of flanking exonic splicing enhancers and involvement of 9G8 as a trans-acting factor. J. Biol. Chem. 276, 7850–7858.

Lejeune, F., Ishigaki, Y., Li, X., Maquat, L.E., 2002. The exon junction complex is detected on CBP80-bound but not eIF4E-bound mRNA in mammalian cells: dynamics of mRNP remodeling. EMBO J. 21, 3536–3545.

Lejeune, F., Li, X., Maquat, L.E., 2003. Nonsense-mediated mRNA decay in mammalian cells involves decapping, deadenylating, and exonucleolytic activities. Mol. Cell 12, 675–687.

Lejeune, F., Ranganathan, A.C., Maquat, L.E., 2004. eIF4G is required for the pioneer round of translation in mammalian cells. Nat. Struct. Mol. Biol. 11, 992–1000.

Lempiainen, H., Halazonetis, T.D., 2009. Emerging common themes in regulation of PIKKs and PI3Ks. EMBO J. 28, 3067–3073.

Leoyklang, P., Suphapeetiporn, K., Srichomthong, C., Tongkobpetch, S., Fietze, S., Dorward, H., Cullinane, A.R., Gahl, W.A., Huizing, M., Shotelersuk, V., 2013. Disorders with similar clinical phenotypes reveal underlying genetic interaction: SATB2 acts as an activator of the UPF3B gene. Hum. Genet. 132, 1383–1393.

Lewis, B.P., Green, R.E., Brenner, S.E., 2003. Evidence for the widespread coupling of alternative splicing and nonsense-mediated mRNA decay in humans. Proc. Natl. Acad. Sci. USA 100, 189–192.

Li, C., Lin, R.I., Lai, M.C., Ouyang, P., Tarn, W.Y., 2003. Nuclear Pnn/DRS protein binds to spliced mRNPs and participates in mRNA processing and export via interaction with RNPS1. Mol. Cell. Biol. 23, 7363–7376.

Linde, L., Boelz, S., Nissim-Rafinia, M., Oren, Y.S., Wilschanski, M., Yaacov, Y., Virgilis, D., Neu-Yilik, G., Kulozik, A.E., Kerem, E., et al., 2007. Nonsense-mediated mRNA decay affects nonsense transcript levels and governs response of cystic fibrosis patients to gentamicin. J. Clin. Invest. 117, 683–692.

Liu, H.X., Cartegni, L., Zhang, M.Q., Krainer, A.R., 2001. A mechanism for exon skipping caused by nonsense or missense mutations in BRCA1 and other genes. Nat. Genet. 27, 55–58.

Liu, J., Valencia-Sanchez, M.A., Hannon, G.J., Parker, R., 2005. MicroRNA-dependent localization of targeted mRNAs to mammalian P-bodies. Nat. Cell Biol. 7, 719–723.

Liu, Y., Bourgeois, C.F., Pang, S., Kudla, M., Dreumont, N., Kister, L., Sun, Y.H., Stevenin, J., Elliott, D.J., 2009. The germ cell nuclear proteins hnRNP G-T and RBMY activate a testis-specific exon. PLoS Genet. 5, e1000707.

Liu, H., Luo, M., Wen, J.K., 2014. mRNA stability in the nucleus. J. Zhejiang Univ. Sci. B 15, 444–454.

Loh, B., Jonas, S., Izaurralde, E., 2013. The SMG5–SMG7 heterodimer directly recruits the CCR4-NOT deadenylase complex to mRNAs containing nonsense codons via interaction with POP2. Genes Dev. 27, 2125–2138.

Losson, R., Lacroute, F., 1979. Interference of nonsense mutations with eukaryotic messenger RNA stability. Proc. Natl. Acad. Sci. USA 76, 5134–5137.

Louhichi, A., Fourati, A., Rebai, A., 2011. IGD: a resource for intronless genes in the human genome. Gene 488, 35–40.

Luo, M.L., Zhou, Z., Magni, K., Christoforides, C., Rappsilber, J., Mann, M., Reed, R., 2001. Pre-mRNA splicing and mRNA export linked by direct interactions between UAP56 and Aly. Nature 413, 644–647.

Lykke-Andersen, J., Bennett, E.J., 2014. Protecting the proteome: eukaryotic cotranslational quality control pathways. J. Cell Biol. 204, 467–476.

Lykke-Andersen, J., Shu, M.D., Steitz, J.A., 2000. Human Upf proteins target an mRNA for nonsense-mediated decay when bound downstream of a termination codon. Cell 103, 1121–1131.

Ma, X.M., Yoon, S.O., Richardson, C.J., Julich, K., Blenis, J., 2008. SKAR links pre-mRNA splicing to mTOR/S6K1-mediated enhanced translation efficiency of spliced mRNAs. Cell 133, 303–313.

Mallick, C.A., Dudley, E.C., Viney, J.L., Owen, M.J., Hayday, A.C., 1993. Rearrangement and diversity of T cell receptor beta chain genes in thymocytes: a critical role for the beta chain in development. Cell 73, 513–519.

Manley, J.L., Krainer, A.R., 2010. A rational nomenclature for serine/arginine-rich protein splicing factors (SR proteins). Genes Dev. 24, 1073–1074.

Maquat, L.E., Kinniburgh, A.J., Rachmilewitz, E.A., Ross, J., 1981. Unstable beta-globin mRNA in mRNA-deficient beta o thalassemia. Cell 27, 543–553.

Markovic, D., Challiss, R.A., 2009. Alternative splicing of G protein-coupled receptors: physiology and pathophysiology. Cell. Mol. Life Sci. 66, 3337–3352.

Markovtsov, V., Nikolic, J.M., Goldman, J.A., Turck, C.W., Chou, M.Y., Black, D.L., 2000. Cooperative assembly of an hnRNP complex induced by a tissue-specific homolog of polypyrimidine tract binding protein. Mol. Cell. Biol. 20, 7463–7479.

Mascarenhas, R., Dougherty, J.A., Schoenberg, D.R., 2013. SMG6 cleavage generates metastable decay intermediates from nonsense-containing beta-globin mRNA. PLoS One 8, e74791.

Matera, A.G., Wang, Z., 2014. A day in the life of the spliceosome. Nat. Rev. Mol. Cell Biol. 15, 108–121.

Matsuoka, S., Ballif, B.A., Smogorzewska, A., McDonald, III, E.R., Hurov, K.E., Luo, J., Bakalarski, C.E., Zhao, Z., Solimini, N., Lerenthal, Y., et al., 2007. ATM and ATR substrate analysis reveals extensive protein networks responsive to DNA damage. Science 316, 1160–1166.

Mazoyer, S., Puget, N., Perrin-Vidoz, L., Lynch, H.T., Serova-Sinilnikova, O.M., Lenoir, G.M., 1998. A BRCA1 nonsense mutation causes exon skipping. Am. J. Hum. Genet. 62, 713–715.

McIlwain, D.R., Pan, Q., Reilly, P.T., Elia, A.J., McCracken, S., Wakeham, A.C., Itie-Youten, A., Blencowe, B.J., Mak, T.W., 2010. Smg1 is required for embryogenesis and regulates diverse genes via alternative splicing coupled to nonsense-mediated mRNA decay. Proc. Natl. Acad. Sci. USA 107, 12186–12191.

Medghalchi, S.M., Frischmeyer, P.A., Mendell, J.T., Kelly, A.G., Lawler, A.M., Dietz, H.C., 2001. Rent1, a trans-effector of nonsense-mediated mRNA decay, is essential for mammalian embryonic viability. Hum. Mol. Genet. 10, 99–105.

Melero, R., Uchiyama, A., Castano, R., Kataoka, N., Kurosawa, H., Ohno, S., Yamashita, A., Llorca, O., 2014. Structures of SMG1-UPFs complexes: SMG1 contributes to regulate UPF2-dependent activation of UPF1 in NMD. Structure 22, 1105–1119.

Mendell, J.T., ap Rhys, C.M., Dietz, H.C., 2002. Separable roles for rent1/hUpf1 in altered splicing and decay of nonsense transcripts. Science 298, 419–422.

Mendell, J.T., Sharifi, N.A., Meyers, J.L., Martinez-Murillo, F., Dietz, H.C., 2004. Nonsense surveillance regulates expression of diverse classes of mammalian transcripts and mutes genomic noise. Nat. Genet. 36, 1073–1078.

Metze, S., Herzog, V.A., Ruepp, M.D., Muhlemann, O., 2013. Comparison of EJC-enhanced and EJC-independent NMD in human cells reveals two partially redundant degradation pathways. RNA 19, 1432–1448.

Mignone, F., Gissi, C., Liuni, S., Pesole, G., 2002. Untranslated regions of mRNAs. Genome Biol. 3, REVIEWS0004.

Miki, Y., Swensen, J., Shattuck-Eidens, D., Futreal, P.A., Harshman, K., Tavtigian, S., Liu, Q., Cochran, C., Bennett, L.M., Ding, W., et al., 1994. A strong candidate for the breast and ovarian cancer susceptibility gene BRCA1. Science 266, 66–71.

Min, H., Chan, R.C., Black, D.L., 1995. The generally expressed hnRNP F is involved in a neural-specific pre-mRNA splicing event. Genes Dev. 9, 2659–2671.

Min, H., Turck, C.W., Nikolic, J.M., Black, D.L., 1997. A new regulatory protein, KSRP, mediates exon inclusion through an intronic splicing enhancer. Genes Dev. 11, 1023–1036.

Murachelli, A.G., Ebert, J., Basquin, C., Le Hir, H., Conti, E., 2012. The structure of the ASAP core complex reveals the existence of a Pinin-containing PSAP complex. Nat. Struct. Mol. Biol. 19, 378–386.

Nagy, E., Maquat, L.E., 1998. A rule for termination-codon position within intron-containing genes: when nonsense affects RNA abundance. Trends Biochem. Sci. 23, 198–199.

Nguyen, L.S., Jolly, L., Shoubridge, C., Chan, W.K., Huang, L., Laumonnier, F., Raynaud, M., Hackett, A., Field, M., Rodriguez, J., et al., 2012. Transcriptome profiling of UPF3B/NMD-deficient lymphoblastoid cells from patients with various forms of intellectual disability. Mol. Psychiatry 17, 1103–1115.

Nguyen, L.S., Kim, H.G., Rosenfeld, J.A., Shen, Y., Gusella, J.F., Lacassie, Y., Layman, L.C., Shaffer, L.G., Gecz, J., 2013. Contribution of copy number variants involving nonsense-mediated mRNA decay pathway genes to neuro-developmental disorders. Hum. Mol. Genet. 22, 1816–1825.

Nicholson, P., Josi, C., Kurosawa, H., Yamashita, A., Muhlemann, O., 2014. A novel phosphorylation-independent interaction between SMG6 and UPF1 is essential for human NMD. Nucleic Acids Res. 42, 9217–9235.

Ohnishi, T., Yamashita, A., Kashima, I., Schell, T., Anders, K.R., Grimson, A., Hachiya, T., Hentze, M.W., Anderson, P., Ohno, S., 2003. Phosphorylation of hUPF1 induces formation of mRNA surveillance complexes containing hSMG-5 and hSMG-7. Mol. Cell 12, 1187–1200.

Okada-Katsuhata, Y., Yamashita, A., Kutsuzawa, K., Izumi, N., Hirahara, F., Ohno, S., 2011. N- and C-terminal Upf1 phosphorylations create binding platforms for SMG-6 and SMG-5:SMG-7 during NMD. Nucleic Acids Res. 40, 1251–1266.

Palacios, I.M., Gatfield, D., St Johnston, D., Izaurralde, E., 2004. An eIF4AIII-containing complex required for mRNA localization and nonsense-mediated mRNA decay. Nature 427, 753–757.

Pan, Q., Shai, O., Lee, L.J., Frey, B.J., Blencowe, B.J., 2008. Deep surveying of alternative splicing complexity in the human transcriptome by high-throughput sequencing. Nat. Genet. 40, 1413–1415.

Pandya-Jones, A., Black, D.L., 2009. Co-transcriptional splicing of constitutive and alternative exons. RNA 15, 1896–1908.

Pandya-Jones, A., Bhatt, D.M., Lin, C.H., Tong, A.J., Smale, S.T., Black, D.L., 2013. Splicing kinetics and transcript release from the chromatin compartment limit the rate of Lipid A-induced gene expression. RNA 19, 811–827.

Park, E., Maquat, L.E., 2013. Staufen-mediated mRNA decay. Wiley Interdiscip. Rev. RNA 4, 423–435.

Pelsy, F., Lacroute, F., 1984. Effect of ochre nonsense mutations on yeast URA1 mRNA stability. Curr. Genet. 8, 277–282.

Pena, V., Liu, S., Bujnicki, J.M., Luhrmann, R., Wahl, M.C., 2007. Structure of a multipartite protein–protein interaction domain in splicing factor prp8 and its link to retinitis pigmentosa. Mol. Cell 25, 615–624.

Pesole, G., Liuni, S., Grillo, G., Licciulli, F., Larizza, A., Makalowski, W., Saccone, C., 2000. UTRdb and UTRsite: specialized databases of sequences and functional elements of 5′ and 3′ untranslated regions of eukaryotic mRNAs. Nucleic Acids Res. 28, 193–196.

Pillai, R.S., Bhattacharyya, S.N., Artus, C.G., Zoller, T., Cougot, N., Basyuk, E., Bertrand, E., Filipowicz, W., 2005. Inhibition of translational initiation by Let-7 microRNA in human cells. Science 309, 1573–1576.

Pinol-Roma, S., Choi, Y.D., Matunis, M.J., Dreyfuss, G., 1988. Immunopurification of heterogeneous nuclear ribonucleoprotein particles reveals an assortment of RNA-binding proteins. Genes Dev. 2, 215–227.

Popp, M.W., Maquat, L.E., 2013. Organizing principles of mammalian nonsense-mediated mRNA decay. Annu. Rev. Genet. 47, 139–165.

Popp, M.W., Maquat, L.E., 2014. The dharma of nonsense-mediated mRNA decay in mammalian cells. Mol. Cells 37, 1–8.

Popp, M.W., Maquat, L.E., 2015. Attenuation of nonsense-mediated mRNA decay facilitates the response to chemotherapeutics. Nat. Commun. 6, 6632.

Porrua, O., Libri, D., 2015. Transcription termination and the control of the transcriptome: why, where and how to stop. Nat. Rev. Mol. Cell Biol. 16, 190–202.

Rebbapragada, I., Lykke-Andersen, J., 2009. Execution of nonsense-mediated mRNA decay: what defines a substrate? Curr. Opin. Cell Biol. 21, 394–402.

Reichenbach, P., Hoss, M., Azzalin, C.M., Nabholz, M., Bucher, P., Lingner, J., 2003. A human homolog of yeast Est1 associates with telomerase and uncaps chromosome ends when overexpressed. Curr. Biol. 13, 568–574.

Reichert, V.L., Le Hir, H., Jurica, M.S., Moore, M.J., 2002. 5′ exon interactions within the human spliceosome establish a framework for exon junction complex structure and assembly. Genes Dev. 16, 2778–2791.

Reznik, B., Lykke-Andersen, J., 2010. Regulated and quality-control mRNA turnover pathways in eukaryotes. Biochem. Soc. Trans. 38, 1506–1510.

Rooke, N., Markovtsov, V., Cagavi, E., Black, D.L., 2003. Roles for SR proteins and hnRNP A1 in the regulation of c-src exon N1. Mol. Cell. Biol. 23, 1874–1884.

Roy, B., Haupt, L.M., Griffiths, L.R., 2013. Review: alternative splicing (AS) of genes as an approach for generating protein complexity. Curr. Genomics 14, 182–194.

Rufener, S.C., Mühlemann, O., 2013. eIF4E-bound mRNPs are substrates for nonsense-mediated mRNA decay in mammalian cells. Nat. Struct. Mol. Biol. 20, 710–717.

Sammeth, M., Foissac, S., Guigo, R., 2008. A general definition and nomenclature for alternative splicing events. PLoS Comput. Biol. 4, e1000147.

Sato, H., Hosoda, N., Maquat, L.E., 2008. Efficiency of the pioneer round of translation affects the cellular site of nonsense-mediated mRNA decay. Mol. Cell 29, 255–262.

Sauliere, J., Murigneux, V., Wang, Z., Marquenet, E., Barbosa, I., Le Tonqueze, O., Audic, Y., Paillard, L., Roest Crollius, H., Le Hir, H., 2012. CLIP-seq of eIF4AIII reveals transcriptome-wide mapping of the human exon junction complex. Nat. Struct. Mol. Biol. 19, 1124–1131.

Schmid, M., Jensen, T.H., 2013. Transcription-associated quality control of mRNP. Biochim. Biophys. Acta 1829, 158–168.

Schmidt, C., Gronborg, M., Deckert, J., Bessonov, S., Conrad, T., Luhrmann, R., Urlaub, H., 2014a. Mass spectrometry-based relative quantification of proteins in precatalytic and catalytically active spliceosomes by metabolic labeling (SILAC), chemical labeling (iTRAQ), and label-free spectral count. RNA 20, 406–420.

Schmidt, S.A., Foley, P.L., Jeong, D.H., Rymarquis, L.A., Doyle, F., Tenenbaum, S.A., Belasco, J.G., Green, P.J., 2014b. Identification of SMG6 cleavage sites and a preferred RNA cleavage motif by global analysis of endogenous NMD targets in human cells. Nucleic Acids Res. 43, 309–323.

Schweingruber, C., Rufener, S.C., Zund, D., Yamashita, A., Muhlemann, O., 2013. Nonsense-mediated mRNA decay—mechanisms of substrate mRNA recognition and degradation in mammalian cells. Biochim. Biophys. Acta 1829, 612–623.

Serin, G., Gersappe, A., Black, J.D., Aronoff, R., Maquat, L.E., 2001. Identification and characterization of human orthologues to Saccharomyces cerevisiae Upf2 protein and Upf3 protein (Caenorhabditis elegans SMG-4). Mol. Cell. Biol. 21, 209–223.

Shabalina, S.A., Ogurtsov, A.Y., Spiridonov, A.N., Novichkov, P.S., Spiridonov, N.A., Koonin, E.V., 2010. Distinct patterns of expression and evolution of intronless and intron-containing mammalian genes. Mol. Biol. Evol. 27, 1745–1749.

Sheth, U., Parker, R., 2003. Decapping and decay of messenger RNA occur in cytoplasmic processing bodies. Science 300, 805–808.

Silva, A.L., Romao, L., 2009. The mammalian nonsense-mediated mRNA decay pathway: to decay or not to decay! Which players make the decision? FEBS Lett. 583, 499–505.

Silva, A.L., Ribeiro, P., Inacio, A., Liebhaber, S.A., Romao, L., 2008. Proximity of the poly(A)-binding protein to a premature termination codon inhibits mammalian nonsense-mediated mRNA decay. RNA 14, 563–576.

Singh, G., Jakob, S., Kleedehn, M.G., Lykke-Andersen, J., 2007. Communication with the exon-junction complex and activation of nonsense-mediated decay by human Upf proteins occur in the cytoplasm. Mol. Cell 27, 780–792.

Singh, G., Rebbapragada, I., Lykke-Andersen, J., 2008. A competition between stimulators and antagonists of Upf complex recruitment governs human nonsense-mediated mRNA decay. PLoS Biol. 6, e111.

Singh, G., Kucukural, A., Cenik, C., Leszyk, J.D., Shaffer, S.A., Weng, Z., Moore, M.J., 2012. The cellular EJC interactome reveals higher-order mRNP structure and an EJC-SR protein nexus. Cell 151, 750–764.

Snow, B.E., Erdmann, N., Cruickshank, J., Goldman, H., Gill, R.M., Robinson, M.O., Harrington, L., 2003. Functional conservation of the telomerase protein Est1p in humans. Curr. Biol. 13, 698–704.

Strasser, K., Hurt, E., 2000. Yra1p, a conserved nuclear RNA-binding protein, interacts directly with Mex67p and is required for mRNA export. EMBO J. 19, 410–420.

Sun, X., Maquat, L.E., 2000. mRNA surveillance in mammalian cells: the relationship between introns and translation termination. RNA 6, 1–8.

Sun, X., Perlick, H.A., Dietz, H.C., Maquat, L.E., 1998. A mutated human homologue to yeast Upf1 protein has a dominant-negative effect on the decay of nonsense-containing mRNAs in mammalian cells. Proc. Natl. Acad. Sci. USA 95, 10009–10014.

Sun, X., Moriarty, P.M., Maquat, L.E., 2000. Nonsense-mediated decay of glutathione peroxidase 1 mRNA in the cytoplasm depends on intron position. EMBO J. 19, 4734–4744.

Sureau, A., Gattoni, R., Dooghe, Y., Stevenin, J., Soret, J., 2001. SC35 autoregulates its expression by promoting splicing events that destabilize its mRNAs. EMBO J. 20, 1785–1796.

Tan, J.T., Kremer, F., Freddi, S., Bell, K.M., Baker, N.L., Lamande, S.R., Bateman, J.F., 2008. Competency for nonsense-mediated reduction in collagen X mRNA is specified by the 3′ UTR and corresponds to the position of mutations in Schmid metaphyseal chondrodysplasia. Am. J. Hum. Genet. 82, 786–793.

Tange, T.O., Shibuya, T., Jurica, M.S., Moore, M.J., 2005. Biochemical analysis of the EJC reveals two new factors and a stable tetrameric protein core. RNA 11, 1869–1883.

Tarpey, P.S., Raymond, F.L., Nguyen, L.S., Rodriguez, J., Hackett, A., Vandeleur, L., Smith, R., Shoubridge, C., Edkins, S., Stevens, C., et al., 2007. Mutations in UPF3B, a member of the nonsense-mediated mRNA decay complex, cause syndromic and nonsyndromic mental retardation. Nat. Genet. 39, 1127–1133.

Thein, S.L., Winichagoon, P., Hesketh, C., Best, S., Fucharoen, S., Wasi, P., Weatherall, D.J., 1990. The molecular basis of beta-thalassemia in Thailand: application to prenatal diagnosis. Am. J. Hum. Genet. 47, 369–375.

Thermann, R., Neu-Yilik, G., Deters, A., Frede, U., Wehr, K., Hagemeier, C., Hentze, M.W., Kulozik, A.E., 1998. Binary specification of nonsense codons by splicing and cytoplasmic translation. EMBO J. 17, 3484–3494.

Thoren, L.A., Norgaard, G.A., Weischenfeldt, J., Waage, J., Jakobsen, J.S., Damgaard, I., Bergstrom, F.C., Blom, A.M., Borup, R., Bisgaard, H.C., et al., 2010. UPF2 is a critical regulator of liver development, function and regeneration. PLoS One 5, e11650.

Toma, K.G., Rebbapragada, I., Durand, S., Lykke-Andersen, J., 2015. Identification of elements in human long 3′ UTRs that inhibit nonsense-mediated decay. RNA 21, 887–897.

Tzschach, A., Grasshoff, U., Beck-Woedl, S., Dufke, C., Bauer, C., Kehrer, M., Evers, C., Moog, U., Oehl-Jaschkowitz, B., Di Donato, N., et al., 2015. Next-generation sequencing in X-linked intellectual disability. Eur. J. Hum. Genet. 23, 1513–1518.

Unterholzner, L., Izaurralde, E., 2004. SMG7 acts as a molecular link between mRNA surveillance and mRNA decay. Mol. Cell 16, 587–596.

Usuki, F., Yamashita, A., Higuchi, I., Ohnishi, T., Shiraishi, T., Osame, M., Ohno, S., 2004. Inhibition of nonsense-mediated mRNA decay rescues the phenotype in Ullrich's disease. Ann. Neurol. 55, 740–744.

van Dijk, E., Cougot, N., Meyer, S., Babajko, S., Wahle, E., Seraphin, B., 2002. Human Dcp2: a catalytically active mRNA decapping enzyme located in specific cytoplasmic structures. EMBO J. 21, 6915–6924.

Vargas, D.Y., Shah, K., Batish, M., Levandoski, M., Sinha, S., Marras, S.A., Schedl, P., Tyagi, S., 2011. Single-molecule imaging of transcriptionally coupled and uncoupled splicing. Cell 147, 1054–1065.

Varsally, W., Brogna, S., 2012. UPF1 involvement in nuclear functions. Biochem. Soc. Trans. 40, 778–783.

Venter, J.C., Adams, M.D., Myers, E.W., Li, P.W., Mural, R.J., Sutton, G.G., Smith, H.O., Yandell, M., Evans, C.A., Holt, R.A., et al., 2001. The sequence of the human genome. Science 291, 1304–1351.

Viegas, M.H., Gehring, N.H., Breit, S., Hentze, M.W., Kulozik, A.E., 2007. The abundance of RNPS1, a protein component of the exon junction complex, can determine the variability in efficiency of the nonsense mediated decay pathway. Nucleic Acids Res. 35, 4542–4551.

Vithana, E.N., Abu-Safieh, L., Allen, M.J., Carey, A., Papaioannou, M., Chakarova, C., Al-Maghtheh, M., Ebenezer, N.D., Willis, C., Moore, A.T., et al., 2001. A human homolog of yeast pre-mRNA splicing gene, PRP31, underlies autosomal dominant retinitis pigmentosa on chromosome 19q13.4 (RP11). Mol. Cell 8, 375–381.

Vitting-Seerup, K., Porse, B.T., Sandelin, A., Waage, J., 2014. spliceR: an R package for classification of alternative splicing and prediction of coding potential from RNA-seq data. BMC Bioinformatics 15, 81.

Wagner, S.D., Berglund, J.A., 2014. Alternative pre-mRNA splicing. Methods Mol. Biol. 1126, 45–54.

Wahl, M.C., Will, C.L., Luhrmann, R., 2009. The spliceosome: design principles of a dynamic RNP machine. Cell 136, 701–718.

Walters, R., Parker, R., 2014. Quality control: is there quality control of localized mRNAs? J. Cell Biol. 204, 863–868.

Wang, P.J., 2004. X chromosomes, retrogenes and their role in male reproduction. Trends Endocrinol. Metab. 15, 79–83.

Wang, J., Gudikote, J.P., Olivas, O.R., Wilkinson, M.F., 2002. Boundary-independent polar nonsense-mediated decay. EMBO Rep. 3, 274–279.

Wang, W., Cajigas, I.J., Peltz, S.W., Wilkinson, M.F., Gonzalez, C.I., 2006. Role for Upf2p phosphorylation in Saccharomyces cerevisiae nonsense-mediated mRNA decay. Mol. Cell. Biol. 26, 3390–3400.

Wang, E.T., Sandberg, R., Luo, S., Khrebtukova, I., Zhang, L., Mayr, C., Kingsmore, S.F., Schroth, G.P., Burge, C.B., 2008. Alternative isoform regulation in human tissue transcriptomes. Nature 456, 470–476.

Welch, E.M., Barton, E.R., Zhuo, J., Tomizawa, Y., Friesen, W.J., Trifillis, P., Paushkin, S., Patel, M., Trotta, C.R., Hwang, S., et al., 2007. PTC124 targets genetic disorders caused by nonsense mutations. Nature 447, 87–91.

Wilkie, S.E., Vaclavik, V., Wu, H., Bujakowska, K., Chakarova, C.F., Bhattacharya, S.S., Warren, M.J., Hunt, D.M., 2008. Disease mechanism for retinitis pigmentosa (RP11) caused by missense mutations in the splicing factor gene PRPF31. Mol. Vis. 14, 683–690.

Wollerton, M.C., Gooding, C., Wagner, E.J., Garcia-Blanco, M.A., Smith, C.W., 2004. Autoregulation of polypyrimidine tract binding protein by alternative splicing leading to nonsense-mediated decay. Mol. Cell 13, 91–100.

Yae, T., Tsuchihashi, K., Ishimoto, T., Motohara, T., Yoshikawa, M., Yoshida, G.J., Wada, T., Masuko, T., Mogushi, K., Tanaka, H., et al., 2012. Alternative splicing of CD44 mRNA by ESRP1 enhances lung colonization of metastatic cancer cell. Nat. Commun. 3, 883.

Yamashita, A., Ohnishi, T., Kashima, I., Taya, Y., Ohno, S., 2001. Human SMG-1, a novel phosphatidylinositol 3-kinase-related protein kinase, associates with components of the mRNA surveillance complex and is involved in the regulation of nonsense-mediated mRNA decay. Genes Dev. 15, 2215–2228.

Yamashita, A., Izumi, N., Kashima, I., Ohnishi, T., Saari, B., Katsuhata, Y., Muramatsu, R., Morita, T., Iwamatsu, A., Hachiya, T., et al., 2009. SMG-8 and SMG-9, two novel subunits of the SMG-1 complex, regulate remodeling of the mRNA surveillance complex during nonsense-mediated mRNA decay. Genes Dev. 23, 1091–1105.

Yepiskoposyan, H., Aeschimann, F., Nilsson, D., Okoniewski, M., Muhlemann, O., 2011. Autoregulation of the nonsense-mediated mRNA decay pathway in human cells. RNA 17, 2108–2118.

You, K.T., Li, L.S., Kim, N.G., Kang, H.J., Koh, K.H., Chwae, Y.J., Kim, K.M., Kim, Y.K., Park, S.M., Jang, S.K., et al., 2007. Selective translational repression of truncated proteins from frameshift mutation-derived mRNAs in tumors. PLoS Biol. 5, e109.

Yu, X., Chini, C.C., He, M., Mer, G., Chen, J., 2003. The BRCT domain is a phosphoprotein binding domain. Science 302, 639–642.

Zahler, A.M., Damgaard, C.K., Kjems, J., Caputi, M., 2004. SC35 and heterogeneous nuclear ribonucleoprotein A/B proteins bind to a juxtaposed exonic splicing enhancer/exonic splicing silencer element to regulate HIV-1 tat exon 2 splicing. J. Biol. Chem. 279, 10077–10084.

Zhai, L.T., Xiang, S., 2014. mRNA quality control at the 5′ end. J. Zhejiang Univ. Sci. B 15, 438–443.

Zhang, Z., Krainer, A.R., 2007. Splicing remodels messenger ribonucleoprotein architecture via eIF4A3-dependent and -independent recruitment of exon junction complex components. Proc. Natl. Acad. Sci. USA 104, 11574–11579.

Zhang, J., Sun, X., Qian, Y., LaDuca, J.P., Maquat, L.E., 1998a. At least one intron is required for the nonsense-mediated decay of triosephosphate isomerase mRNA: a possible link between nuclear splicing and cytoplasmic translation. Mol. Cell. Biol. 18, 5272–5283.

Zhang, J., Sun, X., Qian, Y., Maquat, L.E., 1998b. Intron function in the nonsense-mediated decay of beta-globin mRNA: indications that pre-mRNA splicing in the nucleus can influence mRNA translation in the cytoplasm. RNA 4, 801–815.

Zhang, L., Li, X., Zhao, R., 2013. Structural analyses of the pre-mRNA splicing machinery. Protein Sci. 22, 677–692.

Zhou, Z., Fu, X.D., 2013. Regulation of splicing by SR proteins and SR protein-specific kinases. Chromosoma 122, 191–207.

Zund, D., Gruber, A.R., Zavolan, M., Muhlemann, O., 2013. Translation-dependent displacement of UPF1 from coding sequences causes its enrichment in 3′ UTRs. Nat. Struct. Mol. Biol. 20, 936–943.

CHAPTER 2

Pathologies Susceptible to be Targeted for Nonsense Mutation Therapies

Contents

A genetic disease is defined as a pathology related with a mutation at least in one gene that is responsible for the disorder. The mutation can lead to the total absence of expression of the mutated gene (null mutation or loss of function mutation), to an increase of the wild-type function, or of a new function of the mutated protein (gain of function mutation), or a positive or negative modulation of the expression of the mutated gene. Human genome encodes about 20,000–25,000 genes that can be mutated during DNA replication, DNA repair, or after the insertion or the deletion of transposable elements. However, to date, about 8000 genetic pathologies have been described, suggesting that the other genes either do not have a pathologic effect, or are lethal when they are mutated, explaining why no pathologies are associated with a modification of their expression. It is also possible that the absence of expression of a gene is asymptomatic.

Among the 8000 genetic diseases are found rare genetic diseases, such as cystic fibrosis (CF), Duchenne muscular dystrophy (DMD), hemophilia, or dwarfism, for instance, and frequent genetic diseases such as cancer, metabolic diseases, or neurologic syndromes in which a specific genetic default is often found at the origin of the pathology.

Mutations at the origin of a pathology can be very diversified: point mutation leading to the change of one amino acid to another one (missense mutation), change of one amino acid into a stop codon (nonsense mutation), deletion or insertion of one nucleotide in the open reading frame (ORF) leading to a frameshift mutation -1 or $+1$, respectively. Besides the point mutations, there are also found all the small and large insertions or deletions that are due to replication or transcriptional mistakes, retrotransposon insertion or deletion, any recombination events, or chromosomal migration default. Often, these mutations lead to the generation of a nonfunctional gene product, or to an absence of expression of the mutant gene. However, mutations can also generate the synthesis of a new product with a function deleterious for the cell.

1 RARE DISEASES

A rare disease is a pathology affecting less than 1 person in 1500 in North America, or in 2000 in Europe, for instance, meaning that 30 million Europeans (corresponding to about 8% of the European population) are affected by a rare disease. In 80% of cases, the origin of the pathology is a genetic default and, in 20%, the origin can be a viral or a bacterial infection, or an environmental effect that can induce the pathology. More than 8000 rare diseases have been described and, in most of cases, no treatments are available explaining why rare diseases are also called orphan diseases. Seventy five percent of rare disease patients are children, and the life expectancy does not exceed the age of 5 years in 30% of cases (source: www.eurordis. org). Among these patients with a rare disease due to a genetic cause, about 11% harbor a nonsense mutation in the gene responsible for the pathology (Mort et al., 2008). Correction of the nonsense mutation can potentially apply to about 10% of patients from the 8000 rare diseases, since they are generally monogenic. Three different rare diseases are going to be described later, since they will be used as example in further chapters.

1.1 Duchenne Muscular Dystrophy (DMD)

DMD is a pathology affecting primarily boys, since the gene responsible for the pathology is carried by the short arm of the X-chromosome at the Xp21 locus. The frequency of DMD is one affected boy in 3500 new born boys (Emery, 1993). Mutations affecting the dystrophin gene are responsible for the pathology named Duchenne muscular dystrophy, for the most severe form, and Becker muscular dystrophy (BMD) for the attenuated version of the pathology. BMD patients have a quality of life almost similar to healthy persons and, in some cases, do not even know that they are affected, unlike DMD patients who are subjected to muscle mass decrease leading to the requirement of a wheelchair by the age of 12, and to life expectancy of around 20 years of age. The difference between DMD and BMD is that in DMD no dystrophin

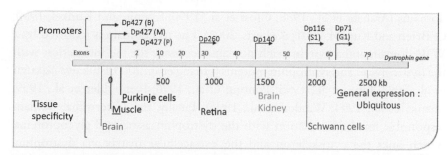

Figure 2.1 *Organization of the dystrophin gene.* On the top are indicated the promoters, with their corresponding tissue specificity in the lower part. The position of some dystrophin exons is shown on the top of the gene, and some distances in kilobases (kb) are provided under the dystrophin gene.

function is detected in these patients when in patients with BMD the function of dystrophin is attenuated often because a mutation decreases the level of dystrophin expression or the level of dystrophin is similar to wild-type but the dystrophin protein is partially functional.

The dystrophin gene is the largest gene in the human genome, covering 2.5 megabases (Coffey et al., 1992; Monaco et al., 1992). This gene encodes 79 exons, and at least seven different promoters have been described that lead to the synthesis of a dystrophin with variable length (Fig. 2.1). Each promoter is expressed in a tissue- or development-specific manner. The main tissues in which dystrophin has been found are in skeletal, cardiac, and smooth muscle, and at lesser extend in neuron cells (Feener et al., 1989; Nudel et al., 1989). Three promoters, located at more than 320 kilobases from the exon 2, drive the expression of the largest version of dystrophin by incorporating a specific exon 1. One promoter expresses dystrophin in the brain (promoter b) (Nudel et al., 1989), another one in muscle and glial cells (promoter m), and finally the last one expresses the biggest isoform of dystrophin in Purkinje cells located in the brain (promoter p) (Gorecki et al., 1992). The dystrophin gene is transcribed into a pre-mRNA subject to alternative splicing, in particular in the last exons, and generates a main 14 kb-long mRNA.

The full size protein synthesized from dystrophin mRNA is a huge protein of 427 kDa, related to the polymerization of 3685 amino acids. The dystrophin protein localizes under the cell membrane in the cytoplasmic side of the sarcolemma, and is a part of the muscle cytoskeleton. Dystrophin has been found in other tissues than muscle, including the brain, where it concentrates at the post-synaptic area, in particular in a subset of GABAergic-synapses (Fritschy et al., 2012; Lidov et al., 1990). Several domains compose dystrophin, starting with an actin-binding site at the N-terminal part (Byers et al., 1989; Fabbrizio et al., 1995; Koenig et al., 1988). The central rod domain is made of 24 spectrin-like triple helical repeats, interrupted by four proline-rich spacer domains (Arahata et al., 1988; Cross et al., 1990; Koenig and Kunkel, 1990; O'Brien and Kunkel, 2001; Roberts, 2001). The rod domain is followed by a WW domain and a cysteine rich domain, necessary for the interaction with the dystroglycan and syntrophin proteins (Huang et al., 2000; Ishikawa-Sakurai et al., 2004; Jung et al., 1995; Ponting et al., 1996; Rentschler et al., 1999; Suzuki et al., 1994; Winder et al., 1995). Finally, the C-terminal domain responsible for the interaction with the dystrophin-associated glycoproteins, which links the cytoskeleton and the extracellular matrix via dystrophin, also includes some putative sites for phosphorylation (Lederfein et al., 1993; Milner et al., 1993; Zubrzycka-Gaarn et al., 1988) (Fig. 2.2).

Besides the full size dystrophin, shorter versions have been described, and called dystrophin protein (Dp), followed by a number corresponding to their molecular weight (Fig. 2.3). These shorter versions of dystrophin

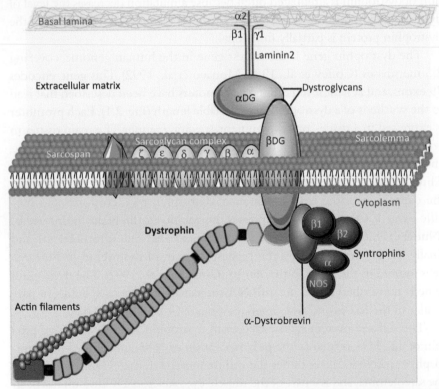

Figure 2.2 *Cellular localization of dystrophin and main partners involved in the muscle cytoskeleton.* Dystrophin localizes under the cell membrane, and interacts with actin filaments on one side, and with β-dystroglycan, α-dystrobrevin, and syntrophin complex on the other side. Thanks to its interactions, dystrophin links the cytoskeleton to the extracellular matrix.

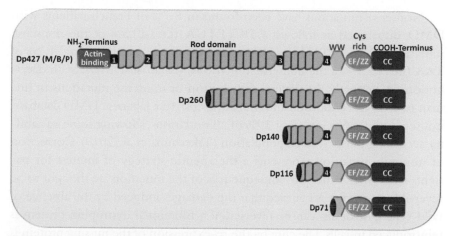

Figure 2.3 *Organization of the functional domains of dystrophin isoforms.* The different domains of dystrophin are indicated, starting by the actin-binding domain, in black, at the N-terminal end of the protein, the rod domain, in blue, composed of 24 repeats of a spectrin-like motif found in the spectrin protein, including four prolin-rich linkers (black boxes named from 1 to 4), the WW protein-protein interaction domain, a cysteine-rich domain composed by a domain EF and a domain ZZ, both also involved in protein-protein interaction, and a coiled coil (cc) domain, also involved in protein–protein interaction. The N-terminal part of the dystrophin is in charge of the interaction with the cytoskeleton, via its actin-binding domain, and the C-terminal part of dystrophin establishes the interaction with the dystrophin-associated proteins (DAP) such as the β-dystroglycan, α-dystrobrevin, and the syntrophins.

are produced thanks to the use of internal promoters that introduce a specific first exon spliced to the exon 30 (Dp260), the exon 45 (Dp140), the exon 56 (Dp116), or the exon 63 (Dp71). Dp260 is mainly expressed in the retina (D'Souza et al., 1995), Dp140 expresses in the fetal brain and, to a lesser extent in the adult brain, in the retina, and in the kidneys (Bardoni et al., 2000; Lidov et al., 1995), Dp116 expresses in adult peripheral nerves (Byers et al., 1993; Labarque et al., 2008), and Dp71—the major dystrophin isoform—in the central nervous system, and is also expressed in various non-muscle tissues (Austin et al., 1995, 2000; Bar et al., 1990; Blake and Kroger, 2000; Ceccarini et al., 2002; Chamberlain et al., 1988; Daoud et al., 2009; Greenberg et al., 1996; Holder et al., 1996; Huard and Tremblay, 1992; Ilarraza-Lomeli et al., 2007; Lederfein et al., 1992; Miyatake et al., 1991; Rapaport et al., 1992). Although Dp71 has a quite ubiquitous expression profile, it is not expressed in skeletal muscle cells. According to the promoter used to generate the dystrophin pre-mRNA, the functional domains included in the protein are different (Fig. 2.3).

Nonsense mutations have been found in 8.7% of French patients with DMD, distributed as follows: 4.3% of TGA (i.e., 49.5% of the nonsense mutations), 2.7% of TAG (i.e., 31% of the nonsense mutations) and 1.7% of TAA (i.e., 19.5% of the nonsense mutations) (database-DMD). Such distribution fits perfectly the general distribution of nonsense mutations in human pathologies (Atkinson and Martin, 1994). In a Japanese DMD database, nonsense mutations represent 19% of all mutations, showing some variability according to the studied population (Takeshima et al., 2010). Correction of nonsense mutation represents a therapeutic strategy of interest for patients with DMD, since the consequences of the mutation are thought to be reversible. Indeed, it is believed that the damages induced by the absence of wild-type dystrophin can be reversed, if a functional dystrophin protein is reintroduced in cells. The cure by the re-expression of the missing protein is expected for most genetic diseases, except the ones affecting the expression of genes required during development, for instance. For that last situation, a treatment would be necessary during embryogenesis and fetal development. Such treatment would have to pass the placenta barrier, and would necessitate the detection of the genetic pathology as early as possible before birth.

Another attractive point for the development of a strategy to correct nonsense mutations in order to cure DMD comes from the fact the C-terminal part of the protein encoded by the exon 70 until the exon 79 is not essential for the function of the protein, at least in muscle cells (Crawford et al., 2000). This means that inhibition of nonsense-mediated mRNA decay (NMD) only to allow the synthesis of a truncated dystrophin protein, when a premature termination codon (PTC) is present in these exons, would be a sufficient therapeutic strategy.

1.2 Cystic Fibrosis (CF)

CF is a recessive pathology related to a dysfunction of the cystic fibrosis transmembrane conductance regulator (CFTR) gene (Kerem et al., 1989; Riordan et al., 1989; Rommens et al., 1989). The CFTR gene is almost 190 kilobases, located on the long arm of the chromosome 7, and it encodes 27 exons. The CFTR mRNA is 6129 nucleotides long, and encodes a protein of 1480 amino acids with a molecular weight around 170 kDa.

The CFTR protein is a member of the ATP binding cassette (ABC) superfamily (ABCC7) that is a family sharing some structural properties, such as the presence of a transmembrane domain (TMD) and a nucleotide-binding domain (NBD), where ATP is recruited and hydrolyzed. More specifically, CFTR is composed of 5 domains: 2 TMDs are present (TMD1 and TMD2), and each one includes 6 transmembrane α-helices; 2 NBDs (NBD1 and

NBD2) and one regulator domain (R) harboring several phosphorylation sites substrates for the cAMP-dependant protein kinases A and C (Fig. 2.4). Both NBD domains bind ATP, thanks to ATP consensus binding sites called Walker A and Walker B (Walker et al., 1982). CFTR is a regulated channel that chloride from inside to outside the cell. The absence of chloride transport leads to an increase of the viscosity of the mucus, in particular in lungs. The function of the lungs is then impaired, and bacteria proliferation is facilitated in such environment found in CF patient lungs.

The CFTR protein is post-translationally modified and, in particular, it is glycosylated before being targeted on the apical plasma membrane of the cell. For that, CFTR is subject to different steps of maturation in different compartments. First, the CFTR protein is synthesized and immediately delivered to the endoplasmic reticulum (ER). In the ER, CFTR is folded

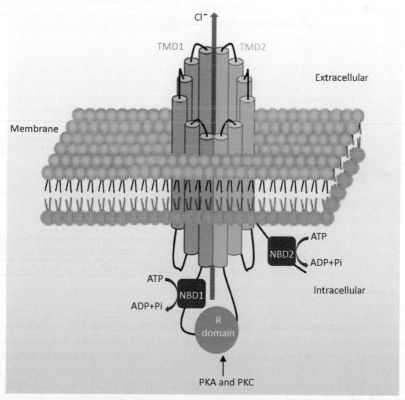

Figure 2.4 *Schematic representation of the CFTR protein.* CFTR is a transmembrane channel transporting chloride from inside to outside of the cell (red arrow). It is composed of two TMDs (TMD1 and TMD2), two NBDs (NBD1 and NBD2), responsible for the catalysis of the ATP, and a regulator (R) domain linking the two identical hemiparts of the protein. The R domain is also substrate for the protein kinases A and C.

and core-glycosylated, in order to give an immature isoform of CFTR, identified by western-blot as band B, with a molecular weight around 145 kDa. Then, CFTR is targeted on the Golgi apparatus, where it is subject to an additional glycosylation in order to generate the CFTR species known as western-blot product band C, of about 170 kDa. This mature form of CFTR then localizes to the apical plasma membrane (Amaral, 2005; Pranke and Sermet-Gaudelus, 2014) (Fig. 2.5).

CF is the most frequent rare disease in the Caucasian population, with one case in about 3000 newborns. It is a recessive and autosomal disorder, with an average life expectancy around 37 years of age (O'Sullivan and Freedman, 2009). CFTR is expressed in most epithelium tissues, explaining

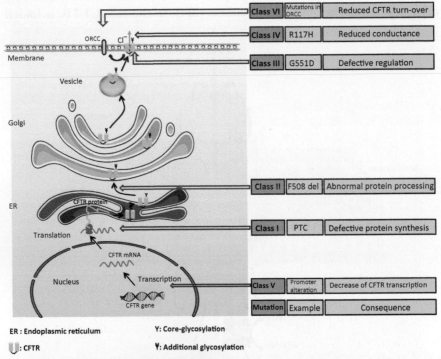

Figure 2.5 *Expression pathway of the CFTR gene, from transcription to the cellular membrane localization.* The CFTR gene is transcribed into CFTR pre-mRNA and, after maturation, into CFTR mRNA before being exported to the cytoplasm, where it is translated into a CFTR protein. The CFTR protein is then glycosylated in the endoplasmic reticulum (ER) before being exported to the Golgi apparatus, where the CFTR protein is subject to another round of glycosylation. CFTR is then fully matured, and can be targeted on the cell membrane. On the right side are mentioned the six main different categories of mutations affecting CFTR (purple box), an example of a mutation type (green box), and the consequence on CFTR (blue box). ORCC: outwardly rectifying chloride channel.

why CF affects various organs, such as lungs, pancreas, intestine, male reproductive organs, or bones, for instance (De Boeck et al., 2006; Fanen et al., 2014; Farrell et al., 2008; Sermet-Gaudelus et al., 2007). Although many organs are impaired in CF, lung dysfunction is the main cause of mortality for CF patients (Fanen et al., 2014; O'Sullivan and Freedman, 2009). To date, 2000 different mutations have been found in CFTR, including the most frequent mutation in CFTR found in 70% of patients—a deletion of the phenylalanine 508 (F508del) that affects the folding of the CFTR protein. Nonsense mutations represent 8.35% of all mutations in the CF mutation database updated in April 2011 (database-CF). In the UMD-CFTR French database, nonsense mutations reach 14% for CF patients, and 5.3% for patients with congenital bilateral absence of vas deferens that is a syndrome present in patients with a mild mutation in CFTR, allowing an expression of about 10% of CFTR, compared to normal. Interestingly, overall in this database, 68% of nonsense mutations are TGA, 24% are TAA, and 8% are TAG (source: UMD-CFTR http://www.umd.be/CFTR/). The distribution of the different nonsense mutations is original, since the less frequent nonsense mutation is TAG, and not TAA unlike what is found in other human pathologies (Atkinson and Martin, 1994). In addition, TGA nonsense mutation shows a large predominance, compared to the proportion found in other human diseases for which TGA nonsense mutation represents around 51% of the nonsense mutations.

Mutations in CFTR have been organized into six different categories (Welsh and Smith, 1993). Class I is composed by all mutations leading to the absence of synthesis of the CFTR protein. For example, nonsense mutations, frameshift mutations, and splicing defaults are found in this category, since the consequence is the introduction of a PTC that promotes the fast CFTR mRNA decay by NMD (Hamosh et al., 1991; Tsui, 1992). Class II regroups mutations impairing the processing of the CFTR protein. It is in this class that the F508del mutation is found, since it interferes with the folding of CFTR that concentrates in the ER, where it is degraded (Cheng et al., 1990; Thomas et al., 1992). In Class III are found all mutations affecting the regulation of the CFTR protein. Mutations located in the NBD, for instance, belong to that category. Indeed, the missense mutation G551D is found in this category, since it reduces strongly the function of CFTR, even though the protein localizes to the apical cell membrane. Class IV is for mutations affecting the conduction of the CFTR channel. Indeed, mutations reduce the ion flow through the channel, or decrease the opening time of CFTR, limiting the exit of chloride ions, and leading to CF. At origin, only four classes have been described, then two additional ones have been proposed, in order to complete the classification of all mutations affecting CFTR (Fanen et al., 2014) (Fig. 2.5). Class V concerns mutations decreasing the production of CFTR mRNA, such as mutations in the CFTR promoter inhibiting the transcription of CFTR, or mutations responsible for the activation of an alternative splicing event leading to the synthesis of nonfunctional or partially functional truncated CFTR protein, for instance (Hinzpeter et al., 2010; Hull et al., 1993). Finally, in class VI are found all mutations increasing the turnover of the functional CFTR protein, or inhibiting the regulation of other ion channels by CFTR, such as outwardly rectifying chloride channel (ORCC), calcium-activated chloride channel, renal outer medullary potassium channel, or the epithelial Na+ channel (Boucher et al., 1986; Gabriel et al., 1993; Haardt et al., 1999; Jovov et al., 1995; Kunzelmann et al., 1997; Mall et al., 2004; Stutts et al., 1997; Toczylowska-Maminska and Dolowy, 2012; Wei et al., 2001; Yoo et al., 2004). Mutations from classes I, II, III, and VI promote an absence or a very low level of CFTR expression, and are associated with a severe clinical phenotype, unlike mutations from classes IV and V, with which a partial CFTR function remains and confers a mild clinical phenotype. It is worth noting that only 5% of CFTR mRNA from the wild-type level is enough to improve the pulmonary clinical phenotype (Ramalho et al., 2002). However, with 10% of CFTR mRNA, compared to healthy persons, people may still develop some syndrome such as congenital bilateral absence of vas deferens in males, leading to azoospermia and sterility (source: UMD-CFTR http://www.umd.be/CFTR/).

Although all the phenotypical characteristics might not be completely reversed, CF is a particularly well adapted pathology for the development of therapeutic strategies correcting nonsense mutations. Indeed, the efficiency of such strategies remains low until now (see chapter: Strategies to Correct Nonsense Mutations; Section 3), but the 5% of CFTR expression sufficient to reverse the main impairment in lung function seems to be accessible, since that level is already reached in different cell models, with various molecules providing a great hope for patients (Du et al., 2009; Manuvakhova et al., 2000).

1.3 Spinal Muscular Atrophy

Spinal muscular atrophy (SMA) is an autosomal recessive disorder caused by mutations in the SMN1 gene encoding the survival motor neuron (SMN) protein for the most common form, but 33 genes have been identified to be potentially the cause of SMA, of which 13 since 2011. SMA affects one newborn in 10,000 and represents the first inherited cause of infant mortality. The proportion of asymptomatic adult carriers is approximatively 1/50

(Sugarman et al., 2012). The consequences of SMA are muscle weakness and atrophy, motor neuron death in the spinal cord, and lower brainstem.

In human, a paralog SMN1 gene exists in the close vicinity of SMN1, and was named SMN2. SMN2 is present in variable copy numbers, from 0 and up to 8 copies that are almost identical to SMN1, with the exception of a silent point mutation transforming a cytosine into thymidine, in the beginning of the exon 7 of the SMN2 gene. The consequence of this mutation is to induce the exon 7 skipping, leading to the synthesis of an unstable truncated SMN protein that is the major SMN isoform in patients with SMA, due to the inhibition of SMN1 expression because of mutations in this gene (Monani et al., 1999). Indeed, this single nucleotide change abolishes an exonic splicing enhancer, and creates an exonic splicing silencer bound by the heterogeneous nuclear ribonucleoprotein A1 (Cartegni and Krainer, 2002; Kashima et al., 2007) (Fig. 2.6). Since a small fraction of SMN2 transcripts includes the exon 7 in the mRNA, the phenotype severity of SMA patients is related to the copy number of the SMN2 gene: the more copies of the SMN2 gene are present, the less

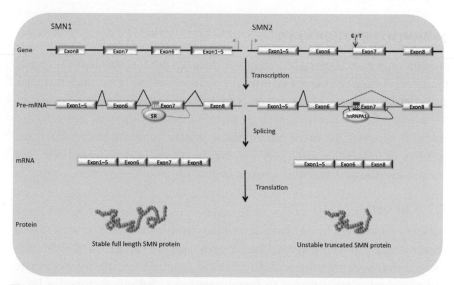

Figure 2.6 *Expression profile of the SMN1 and SMN2 genes.* The SMN1 gene harbors an exonic splicing enhancer in the exon 7, recognized by SR proteins, and leading to the incorporation of the exon 7 in the mRNA, in order to synthesize a functional full length SMN protein. The SMN2 gene is mutated in the exon 7, with a transition of a cytosine into a thymidine. That mutation changes the exonic splicing enhancer into an exonic splicing silencer, leading to the exon 7 skipping during splicing. The translation of such mRNA generates a nonfunctional internally truncated SMN protein.

affected is the patient (Feldkotter et al., 2002; Lefebvre et al., 1997; Prior et al., 2004).

The SMN protein localizes in the nucleus, as well as in the cytoplasm. SMN stably interacts with gemin proteins 2–6 to form the SMN complex (Paushkin et al., 2002) involved in various mechanisms in the cell, including a role in the synthesis, maintenance, and recycling of Sm proteins (Fischer et al., 1997; Meister et al., 2001; Pellizzoni et al., 2002), and in the formation of the U snRNPs (Chari et al., 2008). Indeed, seven Sm proteins (B/B', D1, D2, D3, E, F, and G) are found in the U snRNPs (U1, U2, U4, U5, U11, U12, and U4atac), which are major components of the spliceosome (Lerner and Steitz, 1979; Will and Luhrmann, 2001). Interestingly, the SMN protein, via its function on Sm protein regulations, affects its own expression, since a decrease in the level of SMN protein favors the skipping of the exon 7 of its own pre-mRNA (Jodelka et al., 2010; Ruggiu et al., 2012).

Therapeutic strategies developed to treat SMA concentrated on influencing the splicing of the exon 7 of SMN2, in order to restore the expression of the SMN protein. Some of these strategies will be described in chapter: Strategies to Correct Nonsense Mutations.

2 FREQUENT DISEASES

A pathology is considered as a frequent disease when its incidence is higher than 1 in 3000 newborns. These frequent disorders are often multifactorial, and these factors can be environmental, infectious, or genetic, for example. Such frequent diseases might also require mutations in several genes to develop. To illustrate the involvement of the genetic in frequent diseases, three families of frequent pathologies will be described (cancer, metabolic, and neurologic diseases) via their genetic origin.

2.1 Cancers

Cancers are defined by an anarchic cellular proliferation that forms a tumor within a tissue of an organism. Cancer cells have the capacity to divide endlessly and to become immortal. Tumor cells acquire the capacity to migrate and to colonize other tissues to form metastasis. Multiple factors can be at the origin of a cancer, such as the exposition to radiations (UV, or radioactive, for instance), to stress, to an abnormal hormonal level, to a pathogen, or a cancer-inducer chemical. Mutations in a specific gene can also induce cancer development. For example, mutations in tumor suppressor genes can cause cancer, since those genes are often involved in cell cycle regulation, and in particular in the arrest of cell cycle and, eventually, in the activation of apoptosis. One of the most studied tumor suppressor genes encodes the tumor protein p53 (TP53) that has been found mutated in more than 50% of all human cancers (Carson and Lois, 1995). P53 is a multifunction protein, since it is at least a transcriptional factor, a cell cycle regulator, and an apoptosis activator (Vousden and Prives, 2009). P53 plays a surveillance role on the integrity of the genome. If the genomic DNA is damaged after radiations or chemical exposure, for example, p53 will be involved in the decision to fix the damages by stimulating the expression of the DNA repair factors, or to induce cell death by blocking the cell cycle, and activating the expression of apoptotic genes.

Another well-studied tumor suppressor gene is the breast cancer susceptibility gene 1 (BRCA1). The BRCA1 protein also plays a role in the response to DNA damages, and in particular DNA double strand breaks, by either stimulating the DNA repair or by favoring cell death (Yoshida and Miki, 2004). In case of DNA damage, the BRCA1 protein becomes hyperphosphorylated, and localizes at the DNA replication forks (Scully et al., 1997; Thomas et al., 1997). BRCA1 together with BRCA2 stimulate the double strand DNA repair machinery involving homologous recombination. By interacting with the RNA polymerase II, BRCA1 is also involved in transcription, and stimulates the expression of genes such as p21 or GADD45, proteins involved in cell cycle progression indicating an indirect role of BRCA1 in cell cycle control (Li et al., 2000), consistent with its requirement for the S-phase and the G2/M-phase checkpoints (Cortez et al., 1999).

The phosphatase and tensin homolog (PTEN) is a tumor suppressor acting differently from p53 or BRCA1. PTEN dephosphorylates the phosphatidylinositol 3,4,5 triphosphate (PIP3) into phosphatidylinositol 4,5 diphosphate (PIP2), inhibiting the protein kinase B (Akt) pathway. Indeed, Akt needs the PIP3 to be addressed to the cell membrane, and be activated by phosphorylation. Once Akt is phosphorylated, it can regulate cell survival, and inhibit cell death (Song et al., 2005), explaining how PTEN can be a tumor suppressor by inhibiting the activation of Akt.

The last tumor suppressor gene that will be mentioned is the gene encoding the adenomatous polyposis coli (APC). Mutations in the APC gene increase the risk of developing a colon cancer. APC interacts with several key proteins, such as the beta-catenin, or tubulin forming the microtubules to repress cell division, or to interfere with cell shape and cell mobility (van Es et al., 2001). It is via those interactions that APC interfere with cancer progression.

The proportion of nonsense mutations affecting tumor suppressor genes is variable, according to the gene and the tissue. All the following distributions can be found in the COSMIC library at the following website address: http://cancer.sanger.ac.uk/cosmic/. For instance, nonsense mutations have been found in 7.7% of the 2129 TP53-mutated samples found in the COSMIC library, in 11.4% of the 413 BRCA1-mutated samples, in 15.8% of the 3250 PTEN-mutated samples, and in 41.5% of the 4216 APC-mutated samples. For this latter tumor suppressor gene, the proportion of nonsense mutations falls to 12.5% in the central nervous system, showing the high variability of the distribution of a specific type of mutation, according to the tissue. In any case, the correction of nonsense mutations in cancer would benefit a significant proportion of cancer patients.

Several lines of evidence demonstrate the attractiveness of nonsense mutation correction as a therapeutic approach to cancer. In the case of nonsense mutations affecting a tumor suppressor gene, the correction of the nonsense mutation will rescue the expression of the tumor suppressor gene. In that situation, correction of the nonsense mutation will come as a complementary approach to a therapeutic treatment. Indeed, by rescuing the expression of the mutant tumor suppressor gene, the level of this latter will be reestablished to the physiologic level that is not the level promoting apoptosis, for example. In order to promote cell death, in addition to rescuing the mutant tumor suppressor gene expression, an apoptosis inducing treatment will likely be necessary. This apoptosis treatment will affect the expression level of the tumor suppressor gene, in order to reach the threshold triggering apoptosis. The correction of nonsense mutations is necessary to bring back the normal expression of the mutant tumor suppressor that can then be able to reach the apoptosis triggering level, under apoptosis stimuli.

Correction of nonsense mutations, or precisely inhibition of NMD, can represent another interest in the treatment of cancer, even in the cancer with an origin that is not related to a nonsense mutation. A strong inhibition of NMD leads to the expression of natural substrates of NMD. Among them, some encode proteins involved in apoptosis, including GADD45a, GADD45b, or CDKN1A (Mendell et al., 2004; Popp and Maquat, 2015; Viegas et al., 2007) that will favor the entrance in apoptosis, since these genes are repressed under physiological conditions, and would now become expressed after NMD inhibition. In addition, by dividing continuously, cancer cells increase the probability to get de novo mutations, including nonsense mutations (Campbell et al., 2010; Pleasance et al., 2010). Expressing suddenly a wide panel of truncated proteins might interfere with various physiological processes, and therefore induces apoptosis. It is thus expected to find more mutations in cancer cells than in healthy cells, even inside the same organism, increasing the probability to have more nonsense mutations in cancer cells than in healthy cells. Based on that, inhibition of NMD might be more deleterious in cancer cells than in healthy cells, in side of a cancer patient body.

The third interest in the use of inhibition of NMD, for the treatment of cancer, is that particular approach might be the way to develop an immunotherapy targeting cancer cells. Indeed, inhibition of NMD could lead to the synthesis of many proteins with some abnormal C-terminal end (Fig. 2.7). For example, the translation of mRNA isoforms retaining a PTC-containing intron will generate a truncated protein, with a C-terminal part corresponding to the intronic sequence upstream to the PTC. All these new C-terminal fragments do not exist in healthy cells, and can be part of the

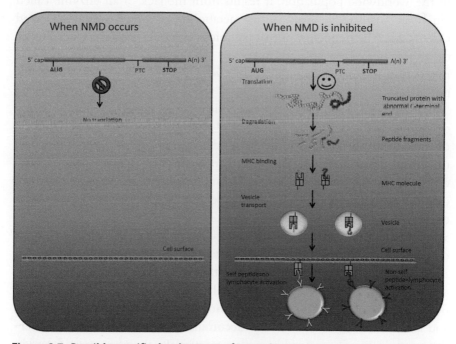

Figure 2.7 *Possible specific development of an antitumor immune response by inhibition of NMD.* When NMD occurs (left panel), PTC containing mRNAs are degraded by NMD, excluding the presentation at the cell surface of tumor specific peptides, unlike when NMD is inhibited, since PTC-containing mRNAs are not quickly degraded, and can be translated into a protein. All the parts of these proteins can be presented at the cell surface, including some C-terminal ends translated from mRNAs, harboring a frameshift mutation or some intronic sequences. Such peptides will not be recognized as self-antigens, and will be responsible for the activation of the immune system.

peptide presented to the cell surface to the immune system. These abnormal C-terminal parts of proteins will then be recognized as nonself, and will activate an immune response. Since cancer cells are more susceptible to accumulate mutations, under NMD inhibition they will present some specific antigens of cancer cells on their cell surface that can promote a specific antitumor immune response (Pastor et al., 2010).

2.2 Metabolic Diseases

Metabolic diseases mean all the pathologies that interfere with the conversion of food into energy. Metabolic diseases can be inherited, known as inborn errors of metabolism, or they can be acquired during the lifetime. The frequency of inherited metabolic diseases is less than 1 in 3000 newborns, making them rare diseases. For example, Gaucher's disease affects 1/60,000 of the worldwide population; it results from the lack of an enzyme called glucocerebrosidase, involved in the metabolism of a fatty substance called cerebroside. The frequency of that pathology is highly variable, according to the ethnicity. For example, it can reach 1/450 people among the Ashkenazi population, making Gaucher's disease a very frequent pathology among this population.

Other well studied inherited metabolic diseases include the Fabrys disorder, with an estimated occurrence of 1/40,000 births, in which a deficiency of alpha-galactosidase A enzyme leads to the accumulation of lipids in various organs, such as kidney or heart, for instance. The type 1 mucopolysaccharidosis, or Hurler syndrome, is another example of inherited metabolic pathology that affects 3–4000 people worldwide. It is a deficiency in L-iduronidase enzyme that leads to the pathology with damage in the heart, lungs, kidney, and central nervous system. L-Iduronidase is responsible for the metabolism of heparin sulfate, and other glycoaminoglycans such as dermatan sulfate. This enzyme localizes in the lysosomes that are organelles in charge of degrading unwanted molecules in the cell (Appelqvist et al., 2013). In case of L-iduronidase impairment, glycoaminoglycans concentrate in lysosomes, and in the extracellular matrix. The accumulation of dermatan sulfate interferes with elastic fiber assembly, explaining partially the phenotype found in Hurler syndrome (Hinek and Wilson, 2000). The main therapeutic strategy for the treatment of metabolic diseases is the enzyme replacement therapy that consists in delivering the missing enzyme to the cell.

Nonsense mutations are found among mutations responsible for metabolic diseases. Nonsense mutation correction seems to be adapted to the treatment of metabolic diseases caused by nonsense mutations, since the level of the enzyme to rescue often does not need to reach 100%. For example, in propionic acidemia mouse model, a rescue of less than 20% of the wild-type level of propionyl-coenzymeA carboxylase, the deficient enzyme in that pathology, is enough to prevent the neonatal lethality (Miyazaki et al., 2001). It has also been shown that, in Hurler syndrome, a low amount—as low as 3% of the L-irudonidase—is sufficient to demonstrate a beneficial effect (Bunge et al., 1998; Keeling et al., 2001). Readthrough molecules have been tested on fibroblasts of patients with Hurler syndrome, and showed their efficiency in inducing expression of L-irudonidase, and in decreasing the concentration of glycoaminoglycans in lyzosomes (Keeling et al., 2001; Wang et al., 2012). The effect of aminoglycosides and non-aminoglycosides, such as ataluren/PTC124, on various inherited metabolic diseases caused by a nonsense mutation have been described positively for peroxisome biogenesis disorders, carnitine palmitoyl transferase deficiency, or branched-chain organic acidurias, including propionic acidemia (Buck et al., 2009; Dranchak et al., 2011; Hein et al., 2004; Helip-Wooley et al., 2002; Keeling et al., 2001; Perez et al., 2012; Sanchez-Alcudia et al., 2012; Sarkar et al., 2011; Tan et al., 2011; Wang et al., 2012).

2.3 Neurologic Disorders

Neurologic diseases group all pathologies affecting the central and/or the peripheral nervous system, including brain, spinal cord, autonomic nervous system, neuromuscular junction, and also muscles. According to the World Health Organization, several hundred millions of people worldwide are affected by a neurologic disorder. More than 600 neurologic pathologies have been reported, with some very common, such as Alzheimer's disease, affecting more than 30 million people worldwide (Querfurth and LaFerla, 2010), Parkinson's disease, epilepsy, affecting more than 50 million people worldwide, or migraine that would affect 10% of the population worldwide.

Nonsense mutations have been found in genes responsible for neurologic disorders. For example, Rett syndrome is a pathology in which nonsense mutations in the methyl CpG binding protein 2 (MECP2) gene, the gene at the origin of the pathology when mutated (Lewis et al., 1992), are found in 27% of Rett syndrome patients (source: RettBASE, http://mecp2.chw. edu.au/cgi-bin/mecp2/views/basic.cgi?form=amino-freq). Unlike other neurologic disorders, Rett syndrome clinical phenotype does not progress after the early childhood step (Naidu et al., 2003). The clinical signs are a developmental arrest, a loss of language acquisition, and a transient autistic-like behavior. The MECP2 gene is carried by the X-chromosome, and encodes for a transcription repressor, and could also be a transcription activator (Chahrour et al., 2008; Cohen et al., 2008). The protein MeCP2 binds to methylated DNA to repress or eventually activate transcription of specific genes, in order to promote the maturation of the central nervous system, and to form synaptic contacts (Akbarian et al., 2001; Jung et al., 2003; Luikenhuis et al., 2004; Shahbazian et al., 2002). Rett syndrome affects girls with a prevalence of 1 in 10000, and rarely boys, likely because it is lethal in males (Samaco et al., 2004; Shahbazian and Zoghbi, 2002). Often, mutations occurring in the MECP2 gene arise de novo. Mutations in the MECP2 gene impair the function of the protein in a dominant negative way, explaining why it would be lethal in affecting boys, and not in girls who would survive thanks to the X-chromosome inactivation that could be the wild-type, or the mutated X-chromosome, leading to a mosaic person (Bienvenu et al., 2002; De Bona et al., 2000; Hoffbuhr et al., 2001, 2002; Zappella et al., 2003). A case study on monozygotic twins carrying mutations in the MECP2 gene illustrates how the process of X-chromosome inactivation can dictate the health status of a Rett syndrome patient. Between these two twins, one girl is asymptomatic, while her sister has a severe Rett syndrome phenotype. A peripheral blood cell analysis demonstrated that the inactivation rate of the X-chromosome is 99:1 in favor of the wild-type X-chromosome for the asymptomatic twin, and a ratio of 40:60 in favor of the mutant X-chromosome for the sister with the severe Rett syndrome phenotype (Hoffbuhr et al., 2001). These data suggest that the severity of the pathology is related to the number of cells expressing the wild-type MeCP2 protein. The dominant effect is only related to the fact that only one allele of the X-chromosome is expressed in females, and does not correspond to a gain of function of the mutant protein. That also explains why it would be lethal for boys, since there is only one copy of the X-chromosome, and if it carries a mutation in MECP2 gene, 100% of cells will not have the functional MECP2 function. Consequently, therapeutic approaches to rescue the clinical phenotype of Rett syndrome patients will be challenging, and the success will depend on the ratio of cells that will respond to the treatment. Indeed, in the example described earlier, 40% of cells expressing the wild-type isoform of the MeCP2 protein are not enough to be asymptomatic. It means that patients expressing wild-type MeCP2, with a ratio of 1:99 in favor of the mutant MeCP2, will be more difficult to be successfully treated, than patients already expressing 40% or 50% of the wild-type protein, suggesting a big difference in the response of putative treatments. In the case of nonsense mutation therapy by the readthrough approach, very efficient compounds are needed since, to date, readthrough efficiency can reach only a few percent of the level of the wild-type protein (see chapter: Strategies to Correct Nonsense Mutations; Section 3).

Another neurologic disorder, epilepsy, has been often related to nonsense mutations. Mutations in various genes lead to the development of epilepsy, such as γ-aminobutyric acid receptor type A (GABA), leucine-rich glioma-inactivated 1 (LGI1), or NaV1.1 neuronal sodium channel alpha-subunit (SCN1A) genes, for instance, illustrating the complexity of epilepsy syndromes. Indeed, several categories of epilepsy have been described, such as the relatively benign generalized epilepsy with febrile seizures plus (GEFS+), the severe myoclonic epilepsy in infancy (SMEI), also known as Dravet syndrome, the borderline SMEI (SMEB), the intractable childhood epilepsy with generalized tonic–clonic seizures (ICEGTCS), some rare cases of familial migraines, the autosomal dominant lateral temporal epilepsy (ADLTE), or the autosomal dominant partial epilepsy with auditory features (ADPEAF). Mutations in one gene can lead to the development of different types of epilepsy, as demonstrated for the SCN1A gene mutations that can cause GEFS+, SMEI, SMEB, ICEGTCS, or familial migraines (Ceulemans et al., 2004; Claes et al., 2001; Escayg et al., 2000; Mulley et al., 2005; Wallace et al., 2001b); in rare cases of familial migraine (Dichgans et al., 2005), LGI1 gene mutations are responsible for ADLTE or ADPEAF (Kalachikov et al., 2002; Morante-Redolat et al., 2002), and mutations in the GABA gene lead to genetic (idiopathic) generalized epilepsy (GGE or IGE), infantile spams (IS), autosomal dominant juvenile myoclonic epilepsy (ADJME), childhood absence epilepsy (CAE), Lennox–Gastaut syndrome (LGS), Dravet syndrome, febrile seizures or GEFS+ (Allen et al., 2013; Audenaert et al., 2006; Baulac et al., 2001; Carvill et al., 2013; Cossette et al., 2002; Delahanty et al., 2011; Dibbens et al., 2004, 2009; Harkin et al., 2002; Hirose, 2014; Ishii et al., 2014; Johnston et al., 2014; Kananura et al., 2002; Lachance-Touchette et al., 2010, 2011; Maljevic et al., 2006; Shi et al., 2010; Sun et al., 2008; Tanaka et al., 2008; Tian et al., 2013; Urak et al., 2006; Wallace et al., 2001a). Interestingly, different mutation types are predominant, according to the category of epilepsy. For example, GEFS+ are often caused by missense mutations (leading to the replacement of one amino acid by another one), while SMEI is mainly due to nonsense mutations or frameshift mutations in the SCN1A gene (Ceulemans et al., 2004). The same as for the MECP2 gene, mutations in SCN1A mainly arise de novo, unlike in the LGI1 gene, in which de novo mutations represent only 2% of mutations (Bisulli et al., 2004; Michelucci et al., 2007).

All the previous data indicate how complex it is to treat one pathology with a unique treatment, rather than to treat the molecular problem at the origin of the pathology. With the development of targeted medicine, we might assist to an inversion of that way of thinking: a treatment will apply to patients having the same molecular dysfunction, and affected by different diseases. The therapeutic unit will not be the pathology any longer, but the molecular event at the origin of the disorder. It is one of these new-targeted therapies that we are going to explore in the next chapters that concern the correction of nonsense mutations.

REFERENCES

Akbarian, S., Chen, R.Z., Gribnau, J., Rasmussen, T.P., Fong, H., Jaenisch, R., Jones, E.G., 2001. Expression pattern of the Rett syndrome gene MeCP2 in primate prefrontal cortex. Neurobiol. Dis. 8, 784–791.

Allen, A.S., Berkovic, S.F., Cossette, P., Delanty, N., Dlugos, D., Eichler, E.E., Epstein, M.P., Glauser, T., Goldstein, D.B., Han, Y., et al., 2013. De novo mutations in epileptic encephalopathies. Nature 501, 217–221.

Amaral, M.D., 2005. Processing of CFTR: traversing the cellular maze—how much CFTR needs to go through to avoid cystic fibrosis? Pediatr. Pulmonol. 39, 479–491.

Appelqvist, H., Waster, P., Kagedal, K., Ollinger, K., 2013. The lysosome: from waste bag to potential therapeutic target. J. Mol. Cell Biol. 5, 214–226.

Arahata, K., Ishiura, S., Ishiguro, T., Tsukahara, T., Suhara, Y., Eguchi, C., Ishihara, T., Nonaka, I., Ozawa, E., Sugita, H., 1988. Immunostaining of skeletal and cardiac muscle surface membrane with antibody against Duchenne muscular dystrophy peptide. Nature 333, 861–863.

Atkinson, J., Martin, R., 1994. Mutations to nonsense codons in human genetic disease: implications for gene therapy by nonsense suppressor tRNAs. Nucleic Acids Res. 22, 1327–1334.

Audenaert, D., Schwartz, E., Claeys, K.G., Claes, L., Deprez, L., Suls, A., Van Dyck, T., Lagae, L., Van Broeckhoven, C., Macdonald, R.L., et al., 2006. A novel GABRG2 mutation associated with febrile seizures. Neurology 67, 687–690.

Austin, R.C., Howard, P.L., D'Souza, V.N., Klamut, H.J., Ray, P.N., 1995. Cloning and characterization of alternatively spliced isoforms of Dp71. Hum. Mol. Genet. 4, 1475–1483.

Austin, R.C., Morris, G.E., Howard, P.L., Klamut, H.J., Ray, P.N., 2000. Expression and synthesis of alternatively spliced variants of Dp71 in adult human brain. Neuromuscul. Disord. 10, 187–193.

Bar, S., Barnea, E., Levy, Z., Neuman, S., Yaffe, D., Nudel, U., 1990. A novel product of the Duchenne muscular dystrophy gene which greatly differs from the known isoforms in its structure and tissue distribution. Biochem. J. 272, 557–560.

Bardoni, A., Felisari, G., Sironi, M., Comi, G., Lai, M., Robotti, M., Bresolin, N., 2000. Loss of Dp140 regulatory sequences is associated with cognitive impairment in dystrophinopathies. Neuromuscul. Disord. 10, 194–199.

Baulac, S., Huberfeld, G., Gourfinkel-An, I., Mitropoulou, G., Beranger, A., Prud'homme, J.F., Baulac, M., Brice, A., Bruzzone, R., LeGuern, E., 2001. First genetic evidence of GABA(A) receptor dysfunction in epilepsy: a mutation in the gamma2-subunit gene. Nat. Genet. 28, 46–48.

Bienvenu, T., Villard, L., De Roux, N., Bourdon, V., Fontes, M., Beldjord, C., Tardieu, M., Jonveaux, P., Chelly, J., 2002. Spectrum of MECP2 mutations in Rett syndrome. Genet. Test. 6, 1–6.

Bisulli, F., Tinuper, P., Scudellaro, E., Naldi, I., Bagattin, A., Avoni, P., Michelucci, R., Nobile, C., 2004. A de novo LGI1 mutation in sporadic partial epilepsy with auditory features. Ann. Neurol. 56, 455–456.

Blake, D.J., Kroger, S., 2000. The neurobiology of Duchenne muscular dystrophy: learning lessons from muscle? Trends Neurosci. 23, 92–99.

Boucher, R.C., Stutts, M.J., Knowles, M.R., Cantley, L., Gatzy, J.T., 1986. Na+ transport in cystic fibrosis respiratory epithelia. Abnormal basal rate and response to adenylate cyclase activation. J. Clin. Invest. 78, 1245–1252.

Buck, N.E., Wood, L., Hu, R., Peters, H.L., 2009. Stop codon read-through of a methylmalonic aciduria mutation. Mol. Genet. Metab. 97, 244–249.

Bunge, S., Clements, P.R., Byers, S., Kleijer, W.J., Brooks, D.A., Hopwood, J.J., 1998. Genotype–phenotype correlations in mucopolysaccharidosis type I using enzyme kinetics, immunoquantification and in vitro turnover studies. Biochim. Biophys. Acta 1407, 249–256.

Byers, T.J., Husain-Chishti, A., Dubreuil, R.R., Branton, D., Goldstein, L.S., 1989. Sequence similarity of the amino-terminal domain of Drosophila beta spectrin to alpha actinin and dystrophin. J. Cell Biol. 109, 1633–1641.

Byers, T.J., Lidov, H.G., Kunkel, L.M., 1993. An alternative dystrophin transcript specific to peripheral nerve. Nat. Genet. 4, 77–81.

Campbell, P.J., Yachida, S., Mudie, L.J., Stephens, P.J., Pleasance, E.D., Stebbings, L.A., Morsberger, L.A., Latimer, C., McLaren, S., Lin, M.L., et al., 2010. The patterns and dynamics of genomic instability in metastatic pancreatic cancer. Nature 467, 1109–1113.

Carson, D.A., Lois, A., 1995. Cancer progression and p53. Lancet 346, 1009–1011.

Cartegni, L., Krainer, A.R., 2002. Disruption of an SF2/ASF-dependent exonic splicing enhancer in SMN2 causes spinal muscular atrophy in the absence of SMN1. Nat. Genet. 30, 377–384.

Carvill, G.L., Heavin, S.B., Yendle, S.C., McMahon, J.M., O'Roak, B.J., Cook, J., Khan, A., Dorschner, M.O., Weaver, M., Calvert, S., et al., 2013. Targeted resequencing in epileptic encephalopathies identifies de novo mutations in CHD2 and SYNGAP1. Nat. Genet. 45, 825–830.

Ceccarini, M., Macioce, P., Panetta, B., Petrucci, T.C., 2002. Expression of dystrophin-associated proteins during neuronal differentiation of P19 embryonal carcinoma cells. Neuromuscul. Disord. 12, 36–48.

Ceulemans, B.P., Claes, L.R., Lagae, L.G., 2004. Clinical correlations of mutations in the SCN1A gene: from febrile seizures to severe myoclonic epilepsy in infancy. Pediatr. Neurol. 30, 236–243.

Chahrour, M., Jung, S.Y., Shaw, C., Zhou, X., Wong, S.T., Qin, J., Zoghbi, H.Y., 2008. MeCP2, a key contributor to neurological disease, activates and represses transcription. Science 320, 1224–1229.

Chamberlain, J.S., Pearlman, J.A., Muzny, D.M., Gibbs, R.A., Ranier, J.E., Caskey, C.T., Reeves, A.A., 1988. Expression of the murine Duchenne muscular dystrophy gene in muscle and brain. Science 239, 1416–1418.

Chari, A., Golas, M.M., Klingenhager, M., Neuenkirchen, N., Sander, B., Englbrecht, C., Sickmann, A., Stark, H., Fischer, U., 2008. An assembly chaperone collaborates with the SMN complex to generate spliceosomal SnRNPs. Cell 135, 497–509.

Cheng, S.H., Gregory, R.J., Marshall, J., Paul, S., Souza, D.W., White, G.A., O'Riordan, C.R., Smith, A.E., 1990. Defective intracellular transport and processing of CFTR is the molecular basis of most cystic fibrosis. Cell 63, 827–834.

Claes, L., Del-Favero, J., Ceulemans, B., Lagae, L., Van Broeckhoven, C., De Jonghe, P., 2001. De novo mutations in the sodium-channel gene SCN1A cause severe myoclonic epilepsy of infancy. Am. J. Hum. Genet. 68, 1327–1332.

Coffey, A.J., Roberts, R.G., Green, E.D., Cole, C.G., Butler, R., Anand, R., Giannelli, F., Bentley, D.R., 1992. Construction of a 2.6-Mb contig in yeast artificial chromosomes spanning the human dystrophin gene using an STS-based approach. Genomics 12, 474–484.

Cohen, S., Zhou, Z., Greenberg, M.E., 2008. Medicine. Activating a repressor. Science 320, 1172–1173.

Cortez, D., Wang, Y., Qin, J., Elledge, S.J., 1999. Requirement of ATM-dependent phosphorylation of brca1 in the DNA damage response to double-strand breaks. Science 286, 1162–1166.

Cossette, P., Liu, L., Brisebois, K., Dong, H., Lortie, A., Vanasse, M., Saint-Hilaire, J.M., Carmant, L., Verner, A., Lu, W.Y., et al., 2002. Mutation of GABRA1 in an autosomal dominant form of juvenile myoclonic epilepsy. Nat. Genet. 31, 184–189.

Crawford, G.E., Faulkner, J.A., Crosbie, R.H., Campbell, K.P., Froehner, S.C., Chamberlain, J.S., 2000. Assembly of the dystrophin-associated protein complex does not require the dystrophin COOH-terminal domain. J. Cell Biol. 150, 1399–1410.

Cross, R.A., Stewart, M., Kendrick-Jones, J., 1990. Structural predictions for the central domain of dystrophin. FEBS Lett. 262, 87–92.

D'Souza, V.N., Nguyen, T.M., Morris, G.E., Karges, W., Pillers, D.A., Ray, P.N., 1995. A novel dystrophin isoform is required for normal retinal electrophysiology. Hum. Mol. Genet. 4, 837–842.

Daoud, F., Angeard, N., Demerre, B., Martie, I., Benyaou, R., Leturcq, F., Cossee, M., Deburgrave, N., Saillour, Y., Tuffery, S., et al., 2009. Analysis of Dp71 contribution in the severity of mental retardation through comparison of Duchenne and Becker patients differing by mutation consequences on Dp71 expression. Hum. Mol. Genet. 18, 3779–3794.

De Boeck, K., Wilschanski, M., Castellani, C., Taylor, C., Cuppens, H., Dodge, J., Sinaasappel, M., 2006. Cystic fibrosis: terminology and diagnostic algorithms. Thorax 61, 627–635.

De Bona, C., Zappella, M., Hayek, G., Meloni, I., Vitelli, F., Bruttini, M., Cusano, R., Loffredo, P., Longo, I., Renieri, A., 2000. Preserved speech variant is allelic of classic Rett syndrome. Eur. J. Hum. Genet. 8, 325–330.

Delahanty, R.J., Kang, J.Q., Brune, C.W., Kistner, E.O., Courchesne, E., Cox, N.J., Cook, Jr, E.H., Macdonald, R.L., Sutcliffe, J.S., 2011. Maternal transmission of a rare GABRB3 signal peptide variant is associated with autism. Mol. Psychiatry 16, 86–96.

Dibbens, L.M., Feng, H.J., Richards, M.C., Harkin, L.A., Hodgson, B.L., Scott, D., Jenkins, M., Petrou, S., Sutherland, G.R., Scheffer, I.E., et al., 2004. GABRD encoding a protein for extra- or peri-synaptic GABAA receptors is a susceptibility locus for generalized epilepsies. Hum. Mol. Genet. 13, 1315–1319.

Dibbens, L.M., Harkin, L.A., Richards, M., Hodgson, B.L., Clarke, A.L., Petrou, S., Scheffer, I.E., Berkovic, S.F., Mulley, J.C., 2009. The role of neuronal GABA(A) receptor subunit mutations in idiopathic generalized epilepsies. Neurosci. Lett. 453, 162–165.

Dichgans, M., Freilinger, T., Eckstein, G., Babini, E., Lorenz-Depiereux, B., Biskup, S., Ferrari, M.D., Herzog, J., van den Maagdenberg, A.M., Pusch, M., et al., 2005. Mutation in the neuronal voltage-gated sodium channel SCN1A in familial hemiplegic migraine. Lancet 366, 371–377.

Dranchak, P.K., Di Pietro, E., Snowden, A., Oesch, N., Braverman, N.E., Steinberg, S.J., Hacia, J.G., 2011. Nonsense suppressor therapies rescue peroxisome lipid metabolism and assembly in cells from patients with specific PEX gene mutations. J. Cell. Biochem. 112 (5), 1250–1258.

Du, L., Damoiseaux, R., Nahas, S., Gao, K., Hu, H., Pollard, J.M., Goldstine, J., Jung, M.E., Henning, S.M., Bertoni, C., et al., 2009. Nonaminoglycoside compounds induce readthrough of nonsense mutations. J. Exp. Med. 206, 2285–2297.

Emery, A.E.H., 1993. Duchenne muscular dystrophy. Oxford University Press, Oxford, UK, p. 392.

Escayg, A., MacDonald, B.T., Meisler, M.H., Baulac, S., Huberfeld, G., An-Gourfinkel, I., Brice, A., LeGuern, E., Moulard, B., Chaigne, D., et al., 2000. Mutations of SCN1A, encoding a neuronal sodium channel, in two families with GEFS+2. Nat. Genet. 24, 343–345.

Fabbrizio, E., Bonet-Kerrache, A., Limas, F., Hugon, G., Mornet, D., 1995. Dystrophin, the protein that promotes membrane resistance. Biochem. Biophys. Res. Commun. 213, 295–301.

Fanen, P., Wohlhuter-Haddad, A., Hinzpeter, A., 2014. Genetics of cystic fibrosis: CFTR mutation classifications toward genotype-based CF therapies. Int. J. Biochem. Cell Biol. 52, 94–102.

Farrell, P.M., Rosenstein, B.J., White, T.B., Accurso, F.J., Castellani, C., Cutting, G.R., Durie, P.R., Legrys, V.A., Massie, J., Parad, R.B., et al., 2008. Guidelines for diagnosis of cystic fibrosis in newborns through older adults: Cystic Fibrosis Foundation consensus report. J. Pediatr. 153, S4–S14.

Feener, C.A., Koenig, M., Kunkel, L.M., 1989. Alternative splicing of human dystrophin mRNA generates isoforms at the carboxy terminus. Nature 338, 509–511.

Feldkotter, M., Schwarzer, V., Wirth, R., Wienker, T.F., Wirth, B., 2002. Quantitative analyses of SMN1 and SMN2 based on real-time lightCycler PCR: fast and highly reliable carrier testing and prediction of severity of spinal muscular atrophy. Am. J. Hum. Genet. 70, 358–368.

Fischer, U., Liu, Q., Dreyfuss, G., 1997. The SMN-SIP1 complex has an essential role in spliceosomal snRNP biogenesis. Cell 90, 1023–1029.

Fritschy, J.M., Panzanelli, P., Tyagarajan, S.K., 2012. Molecular and functional heterogeneity of GABAergic synapses. Cell. Mol. Life Sci. 69, 2485–2499.

Gabriel, S.E., Clarke, L.L., Boucher, R.C., Stutts, M.J., 1993. CFTR and outward rectifying chloride channels are distinct proteins with a regulatory relationship. Nature 363, 263–268.

Gorecki, D.C., Monaco, A.P., Derry, J.M., Walker, A.P., Barnard, E.A., Barnard, P.J., 1992. Expression of four alternative dystrophin transcripts in brain regions regulated by different promoters. Hum. Mol. Genet. 1, 505–510.

Greenberg, D.S., Schatz, Y., Levy, Z., Pizzo, P., Yaffe, D., Nudel, U., 1996. Reduced levels of dystrophin associated proteins in the brains of mice deficient for Dp71. Hum. Mol. Genet. 5, 1299–1303.

Haardt, M., Benharouga, M., Lechardeur, D., Kartner, N., Lukacs, G.L., 1999. C-terminal truncations destabilize the cystic fibrosis transmembrane conductance regulator without impairing its biogenesis. A novel class of mutation. J. Biol. Chem. 274, 21873–21877.

Hamosh, A., Trapnell, B.C., Zeitlin, P.L., Montrose-Rafizadeh, C., Rosenstein, B.J., Crystal, R.G., Cutting, G.R., 1991. Severe deficiency of cystic fibrosis transmembrane conductance regulator messenger RNA carrying nonsense mutations R553X and W1316X in respiratory epithelial cells of patients with cystic fibrosis. J. Clin. Invest. 88, 1880–1885.

Harkin, L.A., Bowser, D.N., Dibbens, L.M., Singh, R., Phillips, F., Wallace, R.H., Richards, M.C., Williams, D.A., Mulley, J.C., Berkovic, S.F., et al., 2002. Truncation of the GABA(A)-receptor gamma2 subunit in a family with generalized epilepsy with febrile seizures plus. Am. J. Hum. Genet. 70, 530–536.

Hein, L.K., Bawden, M., Muller, V.J., Sillence, D., Hopwood, J.J., Brooks, D.A., 2004. Alpha-L-iduronidase premature stop codons and potential read-through in mucopolysaccharidosis type I patients. J. Mol. Biol. 338, 453–462.

Helip-Wooley, A., Park, M.A., Lemons, R.M., Thoene, J.G., 2002. Expression of CTNS alleles: subcellular localization and aminoglycoside correction in vitro. Mol. Genet. Metab. 75, 128–133.

Hinek, A., Wilson, S.E., 2000. Impaired elastogenesis in Hurler disease: dermatan sulfate accumulation linked to deficiency in elastin-binding protein and elastic fiber assembly. Am. J. Pathol. 156, 925–938.

Hinzpeter, A., Aissat, A., Sondo, E., Costa, C., Arous, N., Gameiro, C., Martin, N., Tarze, A., Weiss, L., de Becdelievre, A., et al., 2010. Alternative splicing at a NAGNAG acceptor site as a novel phenotype modifier. PLoS Genet. 6(10), e1001153.

Hirose, S., 2014. Mutant GABA(A) receptor subunits in genetic (idiopathic) epilepsy. Prog. Brain. Res. 213, 55–85.

Hoffbuhr, K., Devaney, J.M., LaFleur, B., Sirianni, N., Scacheri, C., Giron, J., Schuette, J., Innis, J., Marino, M., Philippart, M., et al., 2001. MeCP2 mutations in children with and without the phenotype of Rett syndrome. Neurology 56, 1486–1495.

Hoffbuhr, K.C., Moses, L.M., Jerdonek, M.A., Naidu, S., Hoffman, E.P., 2002. Associations between MeCP2 mutations, X-chromosome inactivation, and phenotype. Ment. Retard Dev. Disabil. Res. Rev. 8, 99–105.

Holder, E., Maeda, M., Bies, R.D., 1996. Expression and regulation of the dystrophin Purkinje promoter in human skeletal muscle, heart, and brain. Hum. Genet. 97, 232–239.

Huang, X., Poy, F., Zhang, R., Joachimiak, A., Sudol, M., Eck, M.J., 2000. Structure of a WW domain containing fragment of dystrophin in complex with beta-dystroglycan. Nat. Struct. Biol. 7, 634–638.

Huard, J., Tremblay, J.P., 1992. Localization of dystrophin in the Purkinje cells of normal mice. Neurosci. Lett. 137, 105–108.

Hull, J., Shackleton, S., Harris, A., 1993. Abnormal mRNA splicing resulting from three different mutations in the CFTR gene. Hum. Mol. Genet. 2, 689–692.

Ilarraza-Lomeli, R., Cisneros-Vega, B., Cervantes-Gomez Mde, L., Mornet, D., Montanez, C., 2007. Dp71, utrophin and beta-dystroglycan expression and distribution in PC12/L6 cell cocultures. Neuroreport 18, 1657–1661.

Ishii, A., Kanaumi, T., Sohda, M., Misumi, Y., Zhang, B., Kakinuma, N., Haga, Y., Watanabe, K., Takeda, S., Okada, M., et al., 2014. Association of nonsense mutation in GABRG2 with abnormal trafficking of GABAA receptors in severe epilepsy. Epilepsy Res. 108, 420–432.

Ishikawa-Sakurai, M., Yoshida, M., Imamura, M., Davies, K.E., Ozawa, E., 2004. ZZ domain is essentially required for the physiological binding of dystrophin and utrophin to beta-dystroglycan. Hum. Mol. Genet. 13, 693–702.

Jodelka, F.M., Ebert, A.D., Duelli, D.M., Hastings, M.L., 2010. A feedback loop regulates splicing of the spinal muscular atrophy-modifying gene, SMN2. Hum. Mol. Genet. 19, 4906–4917.

Johnston, A.J., Kang, J.Q., Shen, W., Pickrell, W.O., Cushion, T.D., Davies, J.S., Baer, K., Mullins, J.G., Hammond, C.L., Chung, S.K., et al., 2014. A novel GABRG2 mutation, p.R136*, in a family with GEFS+ and extended phenotypes. Neurobiol. Dis. 64, 131–141.

Jovov, B., Ismailov, I.I., Berdiev, B.K., Fuller, C.M., Sorscher, E.J., Dedman, J.R., Kaetzel, M.A., Benos, D.J., 1995. Interaction between cystic fibrosis transmembrane conductance regulator and outwardly rectified chloride channels. J. Biol. Chem. 270, 29194–29200.

Jung, D., Yang, B., Meyer, J., Chamberlain, J.S., Campbell, K.P., 1995. Identification and characterization of the dystrophin anchoring site on beta-dystroglycan. J. Biol. Chem. 270, 27305–27310.

Jung, B.P., Jugloff, D.G., Zhang, G., Logan, R., Brown, S., Eubanks, J.H., 2003. The expression of methyl CpG binding factor MeCP2 correlates with cellular differentiation in the developing rat brain and in cultured cells. J. Neurobiol. 55, 86–96.

Kalachikov, S., Evgrafov, O., Ross, B., Winawer, M., Barker-Cummings, C., Martinelli Boneschi, F., Choi, C., Morozov, P., Das, K., Teplitskaya, E., et al., 2002. Mutations in LGI1 cause autosomal-dominant partial epilepsy with auditory features. Nat. Genet. 30, 335–341.

Kananura, C., Haug, K., Sander, T., Runge, U., Gu, W., Hallmann, K., Rebstock, J., Heils, A., Steinlein, O.K., 2002. A splice-site mutation in GABRG2 associated with childhood absence epilepsy and febrile convulsions. Arch. Neurol. 59, 1137–1141.

Kashima, T., Rao, N., David, C.J., Manley, J.L., 2007. hnRNP A1 functions with specificity in repression of SMN2 exon 7 splicing. Hum. Mol. Genet. 16, 3149–3159.

Keeling, K.M., Brooks, D.A., Hopwood, J.J., Li, P., Thompson, J.N., Bedwell, D.M., 2001. Gentamicin-mediated suppression of Hurler syndrome stop mutations restores a low level of alpha-L-iduronidase activity and reduces lysosomal glycosaminoglycan accumulation. Hum. Mol. Genet. 10, 291–299.

Kerem, B., Rommens, J.M., Buchanan, J.A., Markiewicz, D., Cox, T.K., Chakravarti, A., Buchwald, M., Tsui, L.C., 1989. Identification of the cystic fibrosis gene: genetic analysis. Science 245, 1073–1080.

Koenig, M., Kunkel, L.M., 1990. Detailed analysis of the repeat domain of dystrophin reveals four potential hinge segments that may confer flexibility. J. Biol. Chem. 265, 4560–4566.

Koenig, M., Monaco, A.P., Kunkel, L.M., 1988. The complete sequence of dystrophin predicts a rod-shaped cytoskeletal protein. Cell 53, 219–228.

Kunzelmann, K., Mall, M., Briel, M., Hipper, A., Nitschke, R., Ricken, S., Greger, R., 1997. The cystic fibrosis transmembrane conductance regulator attenuates the endogenous Ca^{2+} activated Cl^- conductance of Xenopus oocytes. Pflugers Arch. 435, 178–181.

Labarque, V., Freson, K., Thys, C., Wittevrongel, C., Hoylaerts, M.F., De Vos, R., Goemans, N., Van Geet, C., 2008. Increased Gs signalling in platelets and impaired collagen activation, due to a defect in the dystrophin gene, result in increased blood loss during spinal surgery. Hum. Mol. Genet. 17, 357–366.

Lachance-Touchette, P., Martin, C., Poulin, C., Gravel, M., Carmant, L., Cossette, P., 2010. Screening of GABRB3 in French-Canadian families with idiopathic generalized epilepsy. Epilepsia 51, 1894–1897.

Lachance-Touchette, P., Brown, P., Meloche, C., Kinirons, P., Lapointe, L., Lacasse, H., Lortie, A., Carmant, L., Bedford, F., Bowie, D., et al., 2011. Novel alpha1 and gamma2 GABAA receptor subunit mutations in families with idiopathic generalized epilepsy. Eur. J. Neurosci. 34, 237–249.

Lederfein, D., Levy, Z., Augier, N., Mornet, D., Morris, G., Fuchs, O., Yaffe, D., Nudel, U., 1992. A 71-kilodalton protein is a major product of the Duchenne muscular dystrophy gene in brain and other nonmuscle tissues. Proc. Natl. Acad. Sci. USA 89, 5346–5350.

Lederfein, D., Yaffe, D., Nudel, U., 1993. A housekeeping type promoter, located in the 3' region of the Duchenne muscular dystrophy gene, controls the expression of Dp71, a major product of the gene. Hum. Mol. Genet. 2, 1883–1888.

Lefebvre, S., Burlet, P., Liu, Q., Bertrandy, S., Clermont, O., Munnich, A., Dreyfuss, G., Melki, J., 1997. Correlation between severity and SMN protein level in spinal muscular atrophy. Nat. Genet. 16, 265–269.

Lerner, M.R., Steitz, J.A., 1979. Antibodies to small nuclear RNAs complexed with proteins are produced by patients with systemic lupus erythematosus. Proc. Natl. Acad. Sci. USA 76, 5495–5499.

Lewis, J.D., Meehan, R.R., Henzel, W.J., Maurer-Fogy, I., Jeppesen, P., Klein, F., Bird, A., 1992. Purification, sequence, and cellular localization of a novel chromosomal protein that binds to methylated DNA. Cell 69, 905–914.

Li, S., Ting, N.S., Zheng, L., Chen, P.L., Ziv, Y., Shiloh, Y., Lee, E.Y., Lee, W.H., 2000. Functional link of BRCA1 and ataxia telangiectasia gene product in DNA damage response. Nature 406, 210–215.

Lidov, H.G., Byers, T.J., Watkins, S.C., Kunkel, L.M., 1990. Localization of dystrophin to postsynaptic regions of central nervous system cortical neurons. Nature 348, 725–728.

Lidov, H.G., Selig, S., Kunkel, L.M., 1995. Dp140: a novel 140 kDa CNS transcript from the dystrophin locus. Hum. Mol. Genet. 4, 329–335.

Luikenhuis, S., Giacometti, E., Beard, C.F., Jaenisch, R., 2004. Expression of MeCP2 in postmitotic neurons rescues Rett syndrome in mice. Proc. Natl. Acad. Sci. USA 101, 6033–6038.

Maljevic, S., Krampfl, K., Cobilanschi, J., Tilgen, N., Beyer, S., Weber, Y.G., Schlesinger, F., Ursu, D., Melzer, W., Cossette, P., et al., 2006. A mutation in the GABA(A) receptor alpha(1)-subunit is associated with absence epilepsy. Ann. Neurol. 59, 983–987.

Mall, M., Grubb, B.R., Harkema, J.R., O'Neal, W.K., Boucher, R.C., 2004. Increased airway epithelial Na+ absorption produces cystic fibrosis-like lung disease in mice. Nat. Med. 10, 487–493.

Manuvakhova, M., Keeling, K., Bedwell, D.M., 2000. Aminoglycoside antibiotics mediate context-dependent suppression of termination codons in a mammalian translation system. RNA 6, 1044–1055.

Meister, G., Buhler, D., Pillai, R., Lottspeich, F., Fischer, U., 2001. A multiprotein complex mediates the ATP-dependent assembly of spliceosomal U snRNPs. Nat. Cell Biol. 3, 945–949.

Mendell, J.T., Sharifi, N.A., Meyers, J.L., Martinez-Murillo, F., Dietz, H.C., 2004. Nonsense surveillance regulates expression of diverse classes of mammalian transcripts and mutes genomic noise. Nat. Genet. 36, 1073–1078.

Michelucci, R., Mecarelli, O., Bovo, G., Bisulli, F., Testoni, S., Striano, P., Striano, S., Tinuper, P., Nobile, C., 2007. A de novo LGI1 mutation causing idiopathic partial epilepsy with telephone-induced seizures. Neurology 68, 2150–2151.

Milner, R.E., Busaan, J.L., Holmes, C.F., Wang, J.H., Michalak, M., 1993. Phosphorylation of dystrophin. The carboxyl-terminal region of dystrophin is a substrate for in vitro phosphorylation by p34cdc2 protein kinase. J. Biol. Chem. 268, 21901–21905.

Miyatake, M., Miike, T., Zhao, J.E., Yoshioka, K., Uchino, M., Usuku, G., 1991. Dystrophin: localization and presumed function. Muscle Nerve 14, 113–119.

Miyazaki, T., Ohura, T., Kobayashi, M., Shigematsu, Y., Yamaguchi, S., Suzuki, Y., Hata, I., Aoki, Y., Yang, X., Minjares, C., et al., 2001. Fatal propionic acidemia in mice lacking propionyl-CoA carboxylase and its rescue by postnatal, liver-specific supplementation via a transgene. J. Biol. Chem. 276, 35995–35999.

Monaco, A.P., Walker, A.P., Millwood, I., Larin, Z., Lehrach, H., 1992. A yeast artificial chromosome contig containing the complete Duchenne muscular dystrophy gene. Genomics 12, 465–473.

Monani, U.R., Lorson, C.L., Parsons, D.W., Prior, T.W., Androphy, E.J., Burghes, A.H., McPherson, J.D., 1999. A single nucleotide difference that alters splicing patterns distinguishes the SMA gene SMN1 from the copy gene SMN2. Hum. Mol. Genet. 8, 1177–1183.

Morante-Redolat, J.M., Gorostidi-Pagola, A., Piquer-Sirerol, S., Saenz, A., Poza, J.J., Galan, J., Gesk, S., Sarafidou, T., Mautner, V.F., Binelli, S., et al., 2002. Mutations in the LGI1/Epitempin gene on 10q24 cause autosomal dominant lateral temporal epilepsy. Hum. Mol. Genet. 11, 1119–1128.

Mort, M., Ivanov, D., Cooper, D.N., Chuzhanova, N.A., 2008. A meta-analysis of nonsense mutations causing human genetic disease. Hum. Mutat. 29, 1037–1047.

Mulley, J.C., Scheffer, I.E., Petrou, S., Dibbens, L.M., Berkovic, S.F., Harkin, L.A., 2005. SCN1A mutations and epilepsy. Hum. Mutat. 25, 535–542.

Naidu, S., Bibat, G., Kratz, L., Kelley, R.I., Pevsner, J., Hoffman, E., Cuffari, C., Rohde, C., Blue, M.E., Johnston, M.V., 2003. Clinical variability in Rett syndrome. J. Child Neurol. 18, 662–668.

Nudel, U., Zuk, D., Einat, P., Zeelon, E., Levy, Z., Neuman, S., Yaffe, D., 1989. Duchenne muscular dystrophy gene product is not identical in muscle and brain. Nature 337, 76–78.

O'Brien, K.F., Kunkel, L.M., 2001. Dystrophin and muscular dystrophy: past, present, and future. Mol. Genet. Metab. 74, 75–88.

O'Sullivan, B.P., Freedman, S.D., 2009. Cystic fibrosis. Lancet 373, 1891–1904.

Pastor, F., Kolonias, D., Giangrande, P.H., Gilboa, E., 2010. Induction of tumour immunity by targeted inhibition of nonsense-mediated mRNA decay. Nature 465, 227–230.

Paushkin, S., Gubitz, A.K., Massenet, S., Dreyfuss, G., 2002. The SMN complex, an assemblyosome of ribonucleoproteins. Curr. Opin. Cell Biol. 14, 305–312.

Pellizzoni, L., Yong, J., Dreyfuss, G., 2002. Essential role for the SMN complex in the specificity of snRNP assembly. Science 298, 1775–1779.

Perez, B., Rodriguez-Pombo, P., Ugarte, M., Desviat, L.R., 2012. Readthrough strategies for therapeutic suppression of nonsense mutations in inherited metabolic disease. Mol. Syndromol. 3, 230–236.

Pleasance, E.D., Cheetham, R.K., Stephens, P.J., McBride, D.J., Humphray, S.J., Greenman, C.D., Varela, I., Lin, M.L., Ordonez, G.R., Bignell, G.R., et al., 2010. A comprehensive catalogue of somatic mutations from a human cancer genome. Nature 463, 191–196.

Ponting, C.P., Blake, D.J., Davies, K.E., Kendrick-Jones, J., Winder, S.J., 1996. ZZ and TAZ: new putative zinc fingers in dystrophin and other proteins. Trends Biochem. Sci. 21, 11–13.

Popp, M.W., Maquat, L.E., 2015. Attenuation of nonsense-mediated mRNA decay facilitates the response to chemotherapeutics. Nat. Commun. 6, 6632.

Pranke, I.M., Sermet-Gaudelus, I., 2014. Biosynthesis of cystic fibrosis transmembrane conductance regulator. Int. J. Biochem. Cell Biol. 52, 26–38.

Prior, T.W., Swoboda, K.J., Scott, H.D., Hejmanowski, A.Q., 2004. Homozygous SMN1 deletions in unaffected family members and modification of the phenotype by SMN2. Am. J. Med. Genet. A 130A, 307–310.

Querfurth, H.W., LaFerla, F.M., 2010. Alzheimer's disease. N. Engl. J. Med. 362, 329–344.

Ramalho, A.S., Beck, S., Meyer, M., Penque, D., Cutting, G.R., Amaral, M.D., 2002. Five percent of normal cystic fibrosis transmembrane conductance regulator mRNA ameliorates the severity of pulmonary disease in cystic fibrosis. Am. J. Respir. Cell. Mol Biol. 27, 619–627.

Rapaport, D., Lederfein, D., den Dunnen, J.T., Grootscholten, P.M., Van Ommen, G.J., Fuchs, O., Nudel, U., Yaffe, D., 1992. Characterization and cell type distribution of a novel, major transcript of the Duchenne muscular dystrophy gene. Differentiation 49, 187–193.

Rentschler, S., Linn, H., Deininger, K., Bedford, M.T., Espanel, X., Sudol, M., 1999. The WW domain of dystrophin requires EF-hands region to interact with beta-dystroglycan. Biol. Chem. 380, 431–442.

Riordan, J.R., Rommens, J.M., Kerem, B., Alon, N., Rozmahel, R., Grzelczak, Z., Zielenski, J., Lok, S., Plavsic, N., Chou, J.L., et al., 1989. Identification of the cystic fibrosis gene: cloning and characterization of complementary DNA. Science 245, 1066–1073.

Roberts, R.G., 2001. Dystrophins and dystrobrevins. Genome Biol. 2, REVIEWS3006.

Rommens, J.M., Zengerling-Lentes, S., Kerem, B., Melmer, G., Buchwald, M., Tsui, L.C., 1989. Physical localization of two DNA markers closely linked to the cystic fibrosis locus by pulsed-field gel electrophoresis. Am. J. Hum. Genet. 45, 932–941.

Ruggiu, M., McGovern, V.L., Lotti, F., Saieva, L., Li, D.K., Kariya, S., Monani, U.R., Burghes, A.H., Pellizzoni, L., 2012. A role for SMN exon 7 splicing in the selective vulnerability of motor neurons in spinal muscular atrophy. Mol. Cell. Biol. 32, 126–138.

Samaco, R.C., Nagarajan, R.P., Braunschweig, D., LaSalle, J.M., 2004. Multiple pathways regulate MeCP2 expression in normal brain development and exhibit defects in autism-spectrum disorders. Hum. Mol. Genet. 13, 629–639.

Sanchez-Alcudia, R., Perez, B., Ugarte, M., Desviat, L.R., 2012. Feasibility of nonsense mutation readthrough as a novel therapeutical approach in propionic acidemia. Hum. Mutat. 33, 973–980.

Sarkar, C., Zhang, Z., Mukherjee, A.B., 2011. Stop codon read-through with PTC124 induces palmitoyl-protein thioesterase-1 activity, reduces thioester load and suppresses apoptosis in cultured cells from INCL patients. Mol. Genet. Metab. 104, 338–345.

Scully, R., Chen, J., Ochs, R.L., Keegan, K., Hoekstra, M., Feunteun, J., Livingston, D.M., 1997. Dynamic changes of BRCA1 subnuclear location and phosphorylation state are initiated by DNA damage. Cell 90, 425–435.

Sermet-Gaudelus, I., Souberbielle, J.C., Ruiz, J.C., Vrielynck, S., Heuillon, B., Azhar, I., Caze-nave, A., Lawson-Body, E., Chedevergne, F., Lenoir, G., 2007. Low bone mineral density in young children with cystic fibrosis. Am. J. Respir. Crit. Care Med. 175, 951–957.

Shahbazian, M.D., Zoghbi, H.Y., 2002. Rett syndrome and MeCP2: linking epigenetics and neuronal function. Am. J. Hum. Genet. 71, 1259–1272.

Shahbazian, M.D., Antalffy, B., Armstrong, D.L., Zoghbi, H.Y., 2002. Insight into Rett syn-drome: MeCP2 levels display tissue- and cell-specific differences and correlate with neuronal maturation. Hum. Mol. Genet. 11, 115–124.

Shi, X., Huang, M.C., Ishii, A., Yoshida, S., Okada, M., Morita, K., Nagafuji, H., Yasumoto, S., Kaneko, S., Kojima, T., et al., 2010. Mutational analysis of GABRG2 in a Japanese cohort with childhood epilepsies. J. Hum. Genet. 55, 375–378.

Song, G., Ouyang, G., Bao, S., 2005. The activation of Akt/PKB signaling pathway and cell survival. J. Cell. Mol. Med. 9, 59–71.

Stutts, M.J., Rossier, B.C., Boucher, R.C., 1997. Cystic fibrosis transmembrane conductance regulator inverts protein kinase A-mediated regulation of epithelial sodium channel single channel kinetics. J. Biol. Chem. 272, 14037–14040.

Sugarman, E.A., Nagan, N., Zhu, H., Akmaev, V.R., Zhou, Z., Rohlfs, E.M., Flynn, K., Hen-drickson, B.C., Scholl, T., Sirko-Osadsa, D.A., et al., 2012. Pan-ethnic carrier screen-ing and prenatal diagnosis for spinal muscular atrophy: clinical laboratory analysis of >72,400 specimens. Eur. J. Hum. Genet. 20, 27–32.

Sun, H., Zhang, Y., Liang, J., Liu, X., Ma, X., Wu, H., Xu, K., Qin, J., Qi, Y., Wu, X., 2008. SCN1A, SCN1B, and GABRG2 gene mutation analysis in Chinese families with gen-eralized epilepsy with febrile seizures plus. J. Hum. Genet. 53, 769–774.

Suzuki, A., Yoshida, M., Hayashi, K., Mizuno, Y., Hagiwara, Y., Ozawa, E., 1994. Molecular organization at the glycoprotein-complex-binding site of dystrophin. Three dystrophin-associated proteins bind directly to the carboxy-terminal portion of dystrophin. Eur. J. Biochem. 220, 283–292.

Takeshima, Y., Yagi, M., Okizuka, Y., Awano, H., Zhang, Z., Yamauchi, Y., Nishio, H., Matsuo, M., 2010. Mutation spectrum of the dystrophin gene in 442 Duchenne/Becker muscu-lar dystrophy cases from one Japanese referral center. J. Hum. Genet. 55, 379–388.

Tan, L., Narayan, S.B., Chen, J., Meyers, G.D., Bennett, M.J., 2011. PTC124 improves readthrough and increases enzymatic activity of the CPT1A R160X nonsense muta-tion. J. Inherit. Metab. Dis. 34, 443–447.

Tanaka, M., Olsen, R.W., Medina, M.T., Schwartz, E., Alonso, M.E., Duron, R.M., Castro-Ortega, R., Martinez-Juarez, I.E., Pascual-Castroviejo, I., Machado-Salas, J., et al., 2008. Hyperglycosylation and reduced GABA currents of mutated GABRB3 polypeptide in remitting childhood absence epilepsy. Am. J. Hum. Genet. 82, 1249–1261.

Thomas, P.J., Shenbagamurthi, P., Sondek, J., Hullihen, J.M., Pedersen, P.L., 1992. The cys-tic fibrosis transmembrane conductance regulator. Effects of the most common cystic fibrosis-causing mutation on the secondary structure and stability of a synthetic peptide. J. Biol. Chem. 267, 5727–5730.

Thomas, J.E., Smith, M., Tonkinson, J.L., Rubinfeld, B., Polakis, P., 1997. Induction of phos-phorylation on BRCA1 during the cell cycle and after DNA damage. Cell Growth Differ. 8, 801–809.

Tian, M., Mei, D., Freri, E., Hernandez, C.C., Granata, T., Shen, W., Macdonald, R.L., Guer-rini, R., 2013. Impaired surface alphabetagamma GABA(A) receptor expression in fa-milial epilepsy due to a GABRG2 frameshift mutation. Neurobiol. Dis. 50, 135–141.

Toczylowska-Maminska, R., Dolowy, K., 2012. Ion transporting proteins of human bron-chial epithelium. J. Cell. Biochem. 113, 426–432.

Tsui, L.C., 1992. Mutations and sequence variations detected in the cystic fibrosis transmem-brane conductance regulator (CFTR) gene: a report from the Cystic Fibrosis Genetic Analysis Consortium. Hum. Mutat. 1, 197–203.

Urak, L., Feucht, M., Fathi, N., Hornik, K., Fuchs, K., 2006. A GABRB3 promoter haplotype associated with childhood absence epilepsy impairs transcriptional activity. Hum. Mol. Genet. 15, 2533–2541.

van Es, J.H., Giles, R.H., Clevers, H.C., 2001. The many faces of the tumor suppressor gene APC. Exp. Cell Res. 264, 126–134.

Viegas, M.H., Gehring, N.H., Breit, S., Hentze, M.W., Kulozik, A.E., 2007. The abundance of RNPS1, a protein component of the exon junction complex, can determine the variability in efficiency of the Nonsense Mediated Decay pathway. Nucleic Acids Res. 35, 4542–4551.

Vousden, K.H., Prives, C., 2009. Blinded by the light: the growing complexity of p53. Cell 137, 413–431.

Walker, J.E., Saraste, M., Runswick, M.J., Gay, N.J., 1982. Distantly related sequences in the alpha- and beta-subunits of ATP synthase, myosin, kinases and other ATP-requiring enzymes and a common nucleotide binding fold. EMBO J. 1, 945–951.

Wallace, R.H., Marini, C., Petrou, S., Harkin, L.A., Bowser, D.N., Panchal, R.G., Williams, D.A., Sutherland, G.R., Mulley, J.C., Scheffer, I.E., et al., 2001a. Mutant GABA(A) receptor gamma2-subunit in childhood absence epilepsy and febrile seizures. Nat. Genet. 28, 49–52.

Wallace, R.H., Scheffer, I.E., Barnett, S., Richards, M., Dibbens, L., Desai, R.R., Lerman-Sagie, T., Lev, D., Mazarib, A., Brand, N., et al., 2001b. Neuronal sodium-channel alpha1-subunit mutations in generalized epilepsy with febrile seizures plus. Am. J. Hum. Genet. 68, 859–865.

Wang, D., Belakhov, V., Kandasamy, J., Baasov, T., Li, S.C., Li, Y.T., Bedwell, D.M., Keeling, K.M., 2012. The designer aminoglycoside NB84 significantly reduces glycosaminoglycan accumulation associated with MPS I-H in the Idua-W392X mouse. Mol. Genet. Metab. 105, 116–125.

Wei, L., Vankeerberghen, A., Cuppens, H., Cassiman, J.J., Droogmans, G., Nilius, B., 2001. The C-terminal part of the R-domain, but not the PDZ binding motif, of CFTR is involved in interaction with Ca(2 +)-activated Cl⁻ channels. Pflugers Arch. 442, 280–285.

Welsh, M.J., Smith, A.E., 1993. Molecular mechanisms of CFTR chloride channel dysfunction in cystic fibrosis. Cell 73, 1251–1254.

Will, C.L., Luhrmann, R., 2001. Spliceosomal UsnRNP biogenesis, structure and function. Curr. Opin. Cell Biol. 13, 290–301.

Winder, S.J., Gibson, T.J., Kendrick-Jones, J., 1995. Dystrophin and utrophin: the missing links! FEBS Lett. 369, 27–33.

Yoo, D., Flagg, T.P., Olsen, O., Raghuram, V., Foskett, J.K., Welling, P.A., 2004. Assembly and trafficking of a multiprotein ROMK (Kir 1.1) channel complex by PDZ interactions. J. Biol. Chem. 279, 6863–6873.

Yoshida, K., Miki, Y., 2004. Role of BRCA1 and BRCA2 as regulators of DNA repair, transcription, and cell cycle in response to DNA damage. Cancer Sci. 95, 866–871.

Zappella, M., Meloni, I., Longo, I., Canitano, R., Hayek, G., Rosaia, L., Mari, F., Renieri, A., 2003. Study of MECP2 gene in Rett syndrome variants and autistic girls. Am. J. Med. Genet. B 119B, 102–107.

Zubrzycka-Gaarn, E.E., Bulman, D.E., Karpati, G., Burghes, A.H., Belfall, B., Klamut, H.J., Talbot, J., Hodges, R.S., Ray, P.N., Worton, R.G., 1988. The Duchenne muscular dystrophy gene product is localized in sarcolemma of human skeletal muscle. Nature 333, 466–469.

CHAPTER 3

Strategies to Correct Nonsense Mutations

Contents

Nonsense mutations representing about 10% of the mutations responsible for genetic related pathologies (Mort et al., 2008) attracted the attention of researchers and industries to specifically correct them. Various strategies have been developed in order to correct nonsense mutations. Since nonsense mutations are not purely patient specific and are found in various pathologies, strategies to correct nonsense mutations represent more an example of targeted therapies than a case of personalized medicine. Some of these strategies are nonsense mutation-specific, and some can apply to different categories of mutations. Finally, some will affect the genomic DNA and some will target the RNA only, excluding the ethical question of heredity. Overall, all demonstrate some interesting properties to correct nonsense mutations, but all have weaknesses. A nonexhaustive list of the strategies that can potentially be used to correct nonsense mutations as well as some combinations of strategies are presented in this text.

1 THE EXON SKIPPING

1.1 Principle

Since the identification of introns (also known as intergenic regions) in the second part of the 1970s (Berget et al., 1977; Chow et al., 1977), their removal by splicing reaction has been extensively studied. *Cis-* and *trans-* regulators have been identified and characterized in numerous models, leading to a better understanding of this mechanism (Wang and Burge, 2008; Witten and Ule, 2011). It became therefore possible to interfere with the splicing reaction in order to activate or inhibit it. Several strategies have been developed to inhibit splicing in order to remove the mutation from the mRNA by inducing an exon skipping. Such strategy can therefore apply to the correction of nonsense mutation by removing the PTC-containing exon from the mRNA. Interestingly, the exon skipping strategy also applies to any mutation (insertion, deletion, nonsense, or missense) limited to one or several exon(s), since the skipping can target several exons in order to fix a mutation that affects several exons or to catch back with the original open reading frame (ORF). Indeed, exon skipping opens the possibility to restore an ORF after a deletion of one or several exons, for instance. It becomes a molecular tool to do a surgery on the mRNA in order to fix the consequence of a mutation, without affecting the host genome (Fig. 3.1).

Historically, scientists attempted to prevent the splice site recognition by the spliceosome, using antisense oligonucleotides. The nature of these oligonucleotides evolved from RNA molecules, in the beginning, to reach

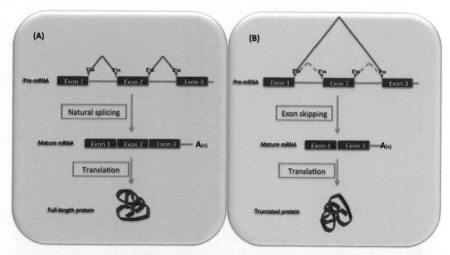

Figure 3.1 *Principle of exon skipping.* (A) An example of constitutive splicing on a pre-mRNA. In that situation, all introns are removed and all exons subsist in the mRNA. (B) An example of an exon skipping occurring on a pre-mRNA. The consequence of the exon 2 skipping of this pre-mRNA is the ligation of the exon 1 with the exon 3, and the absence of the exon 2 from the mRNA. Horizontal lines represent introns and brown boxes symbolize exons. Also, 5′ and 3′ splice site (ss) are indicated.

peptide nucleic acid (PNA) or other modified nucleic acid, in order to improve the strength of the annealing between the oligonucleotide and the RNA target, and to prevent the decay of such oligonucleotide molecules from nuclease activities (Table 3.1) (Aartsma-Rus et al., 2004; Davis et al., 2009; Egholm et al., 1993; Goemans et al., 2011; Karras et al., 2000; Lu et al., 2005; Murray et al., 2012; Summerton and Weller, 1997; Takeshima et al., 2006; Wheeler et al., 2012; Yagi et al., 2004; Yamada et al., 2011).

The idea is to introduce in the cell, a small nucleic acid molecule carrying the antisense sequence of a 5′ or 3′ splice site, in order to prevent the recognition of the target sequence by a *trans* factor. The masked splice site will then not be able to be bound by splicing factors, and the spliceosome will ignore it and use another splice site (Fig. 3.2A).

Another strategy to induce an exon skipping is to tether an inhibitor of splicing close to the targeted splice site, in order to prevent its recognition by the spliceosome. This approach is more complex than the use of antisense oligonucleotide because it requires assembling a molecule that will recognize a specific RNA sequence and a molecule carrying the splicing inhibition activity (Fig. 3.2B). Both strategies present the strong advantage of being sequence-specific, meaning that only the targeted gene will be impacted by the exon skipping, in theory.

Table 3.1 Different types of antisense oligonucleotides used in exon skipping strategy

Drug name	Molecule name	Full name	Structure	References
Negatively charged ON analogues	DNA-PS	DNA-Phosphorothioate		Takeshima et al. (2006)
	ENA	Ethylene-bridged nucleic acids		Yagi et al. (2004)
	LNA	Locked nucleic acid (bridged nucleic acids)		Aartsma-Rus et al. (2004)
	MCE	2'-O-{2-(N-methylcarbamoyl)ethyl} uridine		Yamada et al. (2011)

MOE	2′-O-methoxyethyl			Karras et al. (2000)
MOE gapmers				Wheeler et al. (2012)
OMe-PS (eg, Drisapersen, an OMe-Ps oligonucleotide)	2′-O-methylphosphorothioate			Lu et al. (2005); Goemans et al. (2011)
TcDNA	Tricyclo DNA analogs			Murray et al. (2012)
2′F	2′-Deoxy-2′-fluoronucleotides			Davis et al. (2009)

(Continued)

Table 3.1 Different types of antisense oligonucleotides used in exon skipping strategy (cont.)

Drug name	Molecule name	Full name	Structure	References
Charged neutral peptide nucleic acids and phosphorodiamidate morpholino oligonucleotides	PMO	Phosphorodiamidate morpholino		Summerton and Weller (1997)
	PNA	Peptide nucleic acids		Egholm et al. (1992)

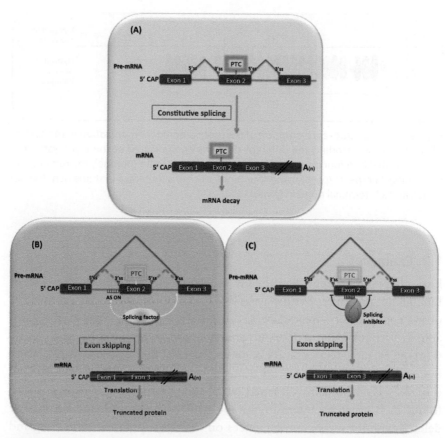

Figure 3.2 *Exon skipping strategies.* (A) A pre-mRNA harboring a PTC in the exon 2 is subject to constitutive splicing resulting in the synthesis of a PTC-containing mRNA that will be degraded by NMD. (B) An antisense oligonucleotide (AS ON) is used to mask the 3′ ss upstream of the exon 2, preventing its recognition by splicing factors that will bind to other 3′ ss. The consequence is the exon 2 skipping and the absence of the PTC in the resulting mRNA. (C) A splicing inhibitor is tethered to the exon 2, thanks to an RNA molecule consisting of a binding sequence for the splicing inhibitor and an antisense sequence to the exon 2. The consequence is the inhibition of the 3′ and 5′ splicing surrounding the exon 2 and the skipping of this exon resulting in the absence of PTC in the mRNA.

As a therapeutic approach, removing an internal part of a protein becomes possible if two parameters are considered: (1) the internally truncated protein part has to be nonessential for the function of the protein and (2) the skipped exon(s) should have a size corresponding to a multiple of three nucleotides, in order to catch back the original ORF when only one exon is planned to be skipped.

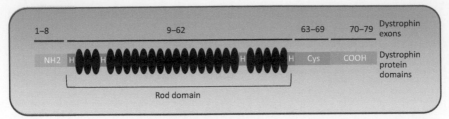

Figure 3.3 *Schematic representation of the protein domain organization of the dystrophin.* The exons encoding the different domains are indicated at the top. Dystrophin can be divided into four domains named N-terminal domain (orange), the rod domain containing 24 repeats (purple) and four hinge domains (green), a cysteine rich domain (blue), and a C-terminal domain (pink).

1.2 Examples

Exon skipping strategy has already reached clinical trials in the case of Duchenne muscular dystrophy (DMD). In this pathology, about 75% of patients could be treated by exon skipping (Aartsma-Rus et al., 2003) and, in particular, about 16% of patients could be targeted by an exon 51 skipping therapy. The exon 51 encodes a part of the central domain of the dystrophin protein called Rod domain, starting from exon 8 to exon 62 (Fig. 3.3). Rod domain is formed by 24 repeats similar to a peptide motif found in β-spectrin. Interestingly, about 60% of the rod domain can be deleted without severe consequences on the dystrophin function (England et al., 1990).

Two strategies have been developed to induce the skipping of the exon 51 or other exons in the rod domain. The first one is to mask the 3′ splice site of an intron in order to induce the skipping of the following exon(s), using antisense oligonucleotide. The second strategy focuses on inhibiting the splicing of a particular exon by tethering a splicing inhibitor on this exon. In order to achieve this, modified U7 snRNA bound by hnRNPA1 was designed in order to anneal with a specific sequence of the target exon (Goyenvalle et al., 2009). Both strategies promoted very encouraging results with the synthesis of internally truncated dystrophin protein in cell culture and animal models, such as mouse or dog (Aartsma-Rus et al., 2003; Barbash et al., 2013; Goyenvalle et al., 2012; Hoogaars et al., 2012; Vulin et al., 2012).

Based on the positive results ex vivo on cell culture, as well as in vivo in mouse and dog models, clinical trials were attempted (for the definition of the clinical trial phases, see Note 3.1).

NOTE 3.1 The Clinical Trial Phases

Clinical trial phase I: This phase is the first study on human and requires a small number of healthy people or patients (between 20 and 80). The toxicity and the tolerance of the drug are evaluated during this phase.

Clinical trial phase II: The aim of this second phase is to determine the minimal efficient dosage. The study is performed on 100–300 voluntary patients who will help to demonstrate some therapeutic benefit, and to identify secondary effects.

Clinical trial phase III: This study measures the efficacy of the drug versus a placebo, or a reference treatment. Several hundred to thousand patients are recruited at this stage. It is the final step before the authorization to put the drug on the market.

Clinical trial phase IV: This phase starts after the introduction of the drug on the market and allows the identification of secondary effects or toxicity after a long period of use.

The number of patients participating in clinical trials can be much lower in the case of the development of treatment on rare diseases, due to the limited number of patients.

Several trials were programmed, such as the one supported by GlaxoSmithKline (GSK) and Prosensa Therapeutics with the molecule Drisapersen, a drug using 2′-O-methyl phosphorothioate antisense oligonucleotide to induce the exon 51 skipping of the dystrophin gene (Clinical trial NCT01803412). At the end of the clinical trial phase II, after 24 weeks of treatment, patients who received Drisapersen succeeded to walk 35.8 m more than patients who received the placebo in the 6-min walk test (6-MWT, see Note 3.2) (Butland et al., 1982). Unfortunately, this drug failed at the clinical trial phase III, since the rescue of the dystrophin function was not significant at the 6-MWT. Another clinical trial, also using antisense approach, has been completed up to the clinical phase II, from Sarepta Therapeutics, with a drug called Eteplirsen that also induces exon skipping of the exon 51 of the dystrophin gene, using a phosphorodiamidate morpholino oligomer (PMO) (Clinical trial NCT 01396239). Results of this clinical trial indicate that patients treated with Eteplirsen were capable of walking about 67 m more than the control group treated with a placebo (Mendell et al., 2013). Patients at the origin of the test were able to walk 200–400 m in the 6-MWT. Interestingly, patients treated with Eteplirsen succeeded to hold their original distance covered during 6 min after 48 weeks of treatment, while patients treated with placebo decreased their performance. This result suggests that the treatment was not able to induce an increase in the muscular mass, but prevents the existing mass to decrease, which is already a very encouraging result.

NOTE 3.2 The 6-Min Walking Test (6-MWT)

This test measures the distance that a patient can cover in 6 min of walk without physical assistance. The patient should do the exercise as fast as possible but it is allowed to slow down or even to stop to rest for a while during the exercise. The test has to be done on a flat ground without any obstacles with a length of at least 25 m without turns.

For a healthy person, the expected distance can be measured by the following formula:

$d = 218 + (5.14 \times \text{size in centimeters}) - (5.32 \times \text{age}) - (1.8 \times \text{weight in kilograms}) + [51.31 \times \text{gender (1 for men and 0 for women)}]$.

As an example, a woman of 45 years of age, 157 cm tall and 50 kg weight, is expected to cover about 696 m.

The exon 51 is not the only exon of the dystrophin eligible for the exon skipping strategy. Indeed, the exon 53 has similar characteristic to exon 51. A clinical trial phase I sponsored by Nippon Shinyaku Pharmaceuticals that started in Jun. 2013 with the drug NS-065/NCNP-01 (NCT02081625) was expected to be completed by Mar. 2015, and target the skipping of the exon 53. The nature of NS-065/NCNP-01 is a morpholino antisense oligonucleotide.

1.3 Weaknesses

Although exon skipping is a highly targeted approach, an interaction of the used antisense sequence with a similar or close sequence present somewhere else in the genome cannot be excluded. It is possible, therefore, that the expression of other genes becomes modified due to the presence of the antisense oligonucleotide or the splicing inhibitor. However, the risk remains limited, since the length of the targeted sequence is more than 20 nucleotides. There is a probability of $1/10^{12}$ to find this sequence, which means that a specific sequence of 20 nucleotides can be found randomly once in a size equivalent of 300 human genomes [the human genome is about 3×10^9 base pairs (bp)].

The second weak point of this strategy is that it will apply only to specific mutations occurring in exon(s) whose size is a multiple factor of three nucleotides, in order to catch back the original ORF. Otherwise several exons will have to be skipped to catch back the original ORF. In addition, the lacking part of the protein has to be nonessential for the function of the protein. These two restrictions strongly reduce the number of patients that can be targeted by this approach.

The third weak point is the control of the number of skipped exons. Indeed, by forcing the spliceosome to ignore one splice site, the aim is often to direct the use of the most proximal splice site and remove an internal part, as small as possible, in order to reduce the impact on the function of the protein. Unfortunately, the use of distal splice sites cannot be prevented, generating many different mRNA isoforms and, eventually, different protein isoforms.

In parallel to preventing the use of a splice site, this strategy should be coupled to a method that favors the use of the splice site, leading to the synthesis of a functional truncated protein. For example, by tethering a splicing activator on the downstream exon of the exon carrying the mutation, it might be possible to favor the use of its splice sites (Fig. 3.4).

The last limitation of this strategy, as a therapeutic approach, is related to the construct delivery into a full organism, or into a specific cell type. This delivery can be done either as a final product (oligonucleotide, ribonucleoprotein) or as an expression vector. In addition, the expression has to be targeted and stable in time. In the case of antisense oligonucleotides strategy, the oligonucleotides have to be provided to the patient constantly, since such molecules will be diluted after cell division and are also subject

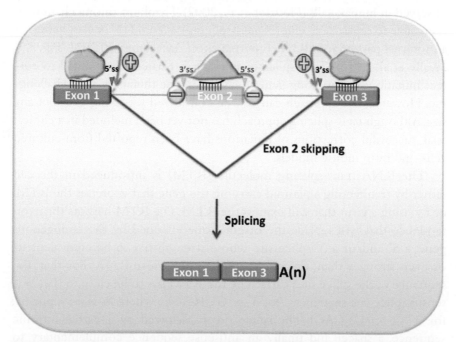

Figure 3.4 *Possible improvement of the exon skipping strategy by concomitant use of splicing inhibitor tether on the exon to skip and splicing activators on the neighbored exons.*

to decay. The way to expose patients, the stability of these molecules in different body fluids, the efficiency of cell penetration, and the amount necessary to get an effect are the biggest challenges for this approach.

In the case of the tethering of splicing inhibitors, the use of modified lentivirus or retrovirus allows the integration of a DNA construct in the genome of the infected cells, leading to a stable expression. The problem is that the integration itself might interfere with the expression of closed genes. In addition, an exogenous DNA element is brought to the genome of patients, with no warranty that this element will not move in the host genome.

2 *TRANS*-SPLICING

2.1 Principle

Trans-splicing is a splicing reaction between two RNA molecules (Fig. 3.5). Basically, the spliceosome uses the 5' splice site from one molecule and the branch point, together with the 3' splice site from another molecule to ligate two exons from two different molecules. This mechanism occurring naturally in eukaryotic cells has been originally identified in trypanosome, and then in flatworms (Davis et al., 1995; Murphy et al., 1986), before being observed in human cells (Flouriot et al., 2002). Thanks to this mechanism, it is possible to replace a 5' end (5' *trans*-splicing), a 3' end (3' *trans*-splicing) or an internal part of an mRNA [internal exon replacement (IER)] (Fig. 3.5) (Wally et al., 2012). *Trans*-splicing approach is particularly adapted to correct mutations in very long genes for which gene therapy is very challenging. However, this approach can also be considered for fixing genes of any size. Although *trans*-splicing approach has not yet been included in a clinical trial, successful corrections of mutations have been reported from cultured cells, and from mouse models.

The RNA trans-splicing molecule (RTM) is introduced in the cell either by transfecting a plasmid carrying the gene that expresses the RTM, or by using a virus that will express the RTM. The RTM harbors the exon sequence that will replace the original one encoded by the endogenous gene, a 5' and/or a 3' splice site whose strength has to be equivalent to, or even stronger than, the one carried by the pre-mRNA. For that, the sequence of the splice sites carried by the RTM has to be close to the consensus splice site sequences (for 5' ss: GURAGU, where R is for a purine; for 3' ss: $(C/U)_n CAG$). The splice site is followed by a partial intronic sequence, a spacer and finally an antisense sequence complementary to a portion of the intron to be *trans*-spliced (Fig. 3.5). This last part of the

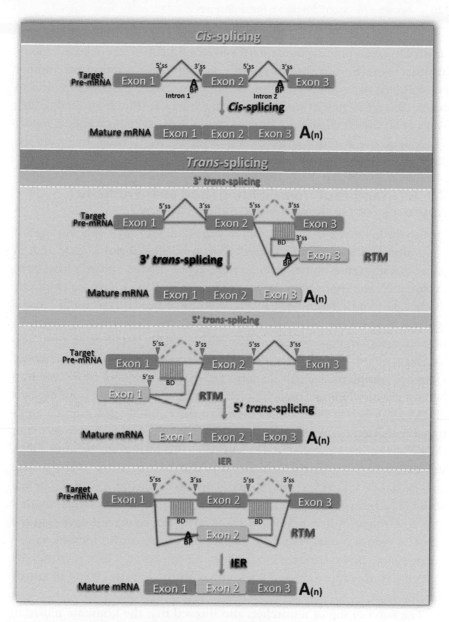

Figure 3.5 *The different categories of* trans-*splicing*. At the top, an example of *cis*-splicing is shown. The second panel from the top explains the 3' *trans*-splicing in which a downstream exon is brought by an RNA *trans*-splicing molecule (RTM). The splicing of the intron 2 is performed between the 5' ss of the pre-mRNA and the 3' ss of the RTM. The third panel from the top represents an example of 5' *trans*-splicing, in which an upstream exon is brought by the RTM. The splicing reaction at the intron 1 is done between the 5' ss of the RTM and the 3' ss of the intron 1 of the pre-mRNA. The bottom panel represents an example of IER for which the splicing of the intron 1 is performed between the 5' ss of the intron 1 of the pre-mRNA and the 3' ss of the RTM, and the splicing of the intron 2 is done between the 5' ss of the RTM and the 3' ss of the intron 2 of the pre-mRNA. The branch point (BP) of each intronic sequence is symbolized by an "A."

construct, called binding domain (BD), is around 100–150 nucleotides long, in order to minimize the nonspecific *trans*-splicing reaction on another pre-mRNA, or at another intron in the same target pre-mRNA, and to ensure the efficiency of the *trans*-splicing reaction (Puttaraju et al., 2001; Walls et al., 2008). Finally, in the case of a 3′ *trans*-splicing reaction, a poly adenylation site is added at the 3′ end of the RTM, in order to ensure the correct processing of the *trans*-spliced mRNA.

2.2 An Example of *Trans*-Splicing Used as Therapeutic Approach for Duchenne Muscular Dystrophy

Due to the giant size of the dystrophin gene (2.5 Mb), *trans*-splicing appears as an attractive approach to correct mutations affecting this gene and, in particular, nonsense mutations. The three types of *trans*-splicing have been explored on the dystrophin model in vitro, as well as in vivo, since mouse models exist for that pathology. In 2013, a study showed that by using 3′ *trans*-splicing, it is possible to correct a nonsense mutation present in the exon 23 of the dystrophin gene found in *mdx* mouse model (Lorain et al., 2013). To do that, they used an adeno-associated virus construct carrying an antisense sequence complementary to the intron 22, a spacer sequence followed by a hemi intron, including a branch point and a 3′ splice site, and finishing by a cDNA sequence of the exon 23 ligated to the exon 59, generating a functional truncated dystrophin (Fig. 3.6). The AAV constructs were injected in the tibialis anterior muscle for 4 weeks. After this treatment, genomic DNA was extracted to verify the integration of the AAV construct. RNAs were also extracted to analyze the efficiency of *trans*-splicing; this had reached about 30%. Interestingly, the same construct can be used to correct, another nonsense mutation located in the exon 53 of the dystrophin gene, in the *mdx4Cv* mouse model (Fig. 3.6). Indeed, several mutations located on different exons can be corrected with the same construct, which increases the potential of this approach and especially with the prospect of an industrial development.

The same group of researchers also showed that the nonsense mutation present in the exon 23 of the dystrophin gene can be corrected in *mdx* mice by using an IER *trans*-splicing strategy (Lorain et al., 2010). The RTM they used started from the 5′ end by an antisense sequence complementary to the intron 22, a spacer, then a hemi intron containing a branch point and a 3′ splice site, the exon 24, a 5′ splice site and a hemi intron sequence, a spacer, an antisense sequence complementary to the intron 23, and finally, a polyadenylation signal. Intronic splicing enhancer sequences were added

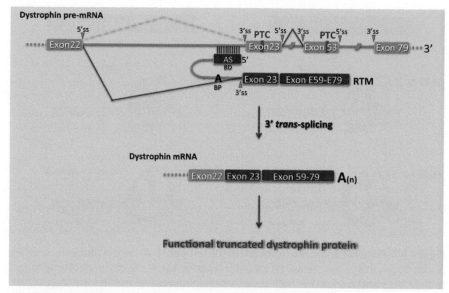

Figure 3.6 *Correction of nonsense mutation in the dystrophin gene in the exon 23 or 53 using a unique RTM.* The *trans*-splicing reaction occurred between the 5′ ss of the dystrophin pre-mRNA and the 3′ ss of the RTM. The RTM consists of a binding domain (BD) of about 150 nucleotides, the BP symbolized by a "A," the 3′ ss and the exon 23 followed by the dystrophin exons from 59 to 79. The polyadenylation site was also included in the RTM. The protein encoded by the *trans*-spliced mRNA is functional, even with the internal truncation.

to each hemi intron sequence in order to improve the efficiency of *trans*-splicing about ninefold, allowing it to reach 45% of *trans*-spliced dystrophin mRNA.

2.3 Weaknesses

The major weak point of this approach is the efficiency of *trans*-splicing that can reach 45% but may be not sufficient to cure the consequences of a mutation. Another challenge is the toxicity observed coming from the RTM. Indeed, translation can occur directly on the RTM, and it leads to the production of unwanted peptides with a putative risk of toxicity. The production of these proteins from the RTM can be very complex to identify, since *cis*-splicing can also occur on the RTM, generating more possibilities to create some ORFs (Monjaret et al., 2014). This approach has not yet reached the clinical trial stage due to concerns about its safety, but it still represents a very attractive approach to fix any type of mutations, particularly in big genes.

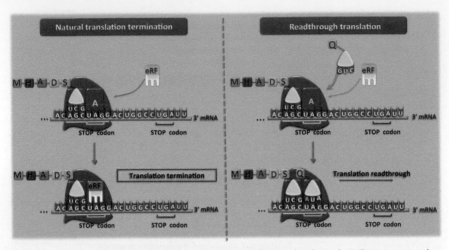

Figure 3.7 *Principle of the readthrough.* In the left panel, when the ribosome reaches a stop codon (UAG), the release factor (eRF) enters in the A site of the ribosome to terminate the translation. In the right panel, when the ribosome reaches the stop codon, a competition occurs between the release factor and a tRNA that can enter the A site to recognize the stop codon under readthrough circumstances. The translation can then go on until the next stop codon.

3 PTC-READTHROUGH

3.1 Principle

Readthrough is a natural mechanism that consists in the incorporation of an amino acid at the PTC position (Fig. 3.7). It could be considered as a mistake done by the translation machinery since the stop codon is not recognized by the translation termination machinery. Readthrough of stop codons has been found to occur naturally on particular mRNAs and in various species. Indeed, about 5% of stop codons are reassigned in *Drosophila* (Jungreis et al., 2011). In human, the identification of natural readthrough is under investigation and only five genes have been clearly demonstrated to use readthrough to modulate their expression (SACM1L, OPRK1, OPRL1, BRI3BP, and MYELIN P0) and at least three others are predicted to be subject to readthrough on their stop codon (ACP2, MAPK10, and AQP4) (Loughran et al., 2014; Yamaguchi et al., 2012). An increasing number of human genes supporting readthrough on their normal termination codon is expected to be identified in a close future, since specific nucleotide context has been identified to favor readthrough, and such sequence can be found in many genes. As a matter of fact, the 5′ and the 3′ sequences influence the rate of readthrough, and the immediate stop codon 3′ sequences CUAG or at a more degenerated level, the sequence CARYYA (R for purine and Y for pyrimidine) are found around stop codons subject to readthrough, in most of cases (Loughran et al., 2014). Until now, natural readthrough has been found to occur on UGA stop codon, likely because this stop codon is the leakiest. However, readthrough can also occur on UAG or UAA stop codon with a lower efficiency than on the UGA stop codon as demonstrated in cell culture (Loughran et al., 2014).

In the case of a nonsense mutation, readthrough of the PTC represents a very interesting therapeutic strategy, since the final product would be the synthesis of the full-length protein, with a maximum of one amino acid different from the wild-type protein. Indeed, the incorporated amino acid at the PTC position could be the same as the one present in the wild-type protein, or a different one. This means that, in most of cases, the protein will be functional, except if the amino acid incorporated at the PTC position is crucial for the function of the protein and not compatible with this function. The incorporated amino acid depends on the identity of the stop codon that is readthrough. Indeed, a UAA or UAG stop codon will favor the incorporation of a glutamine at the PTC position, while a UGA stop codon will promote the incorporation of an arginine, a tryptophane, or a cysteine (Feng et al., 1990). A recent study in yeast refines the incorporation rate of amino acid at the PTC position as follows: there is a 54% chance to incorporate a tyrosine, a 44% chance to incorporate a glutamine, and a 2% chance to incorporate a lysine at a UAA stop codon; for the UAG stop codon, tyrosine will be incorporated in 92% of cases, glutamine in 5% of cases, and lysine in 3% of cases. Finally, for the UGA stop codon, tryptophane is the major incorporated amino acid (82% of cases), then the cysteine (14–17% of cases) followed by arginine (1–4% of cases) (Blanchet et al., 2014). In the absence of a dedicated analysis, it cannot be excluded that the incorporation rate of amino acids could be slightly different in other species and could depend on the molecule used to activate readthrough.

In the development of nonsense suppression therapies by readthrough, there are two major parameters that need to be taken in account: (1) the identity of the stop codon, since, in the case of many molecules activating readthrough, UGA has been shown to be the easiest stop codon to be readthrough, and UAA the most difficult one and (2) the nucleotide context of the stop codon. We saw that, in the case of a natural readthrough on the physiological stop codon, the CUAG sequence immediately downstream of the stop codon favors readthrough (Loughran et al., 2014). The influence of the nucleotide context in the vicinity of the PTC has been studied, and it

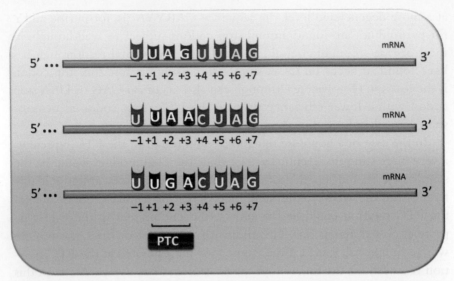

Figure 3.8 *Nucleotide sequences promoting efficient readthrough according to the stop codon.* The number under the mRNA corresponds to the position relative to the stop codon.

was demonstrated that the presence of an uridine immediately upstream of the PTC and a cytosine downstream at the position +4 promote the highest rate of readthrough induced by gentamicin, making the sequence "U stop C" the best candidate for readthrough therapy (Floquet et al., 2012). However, the nucleotide context influences differently the readthrough according to the identity of the stop codon (Manuvakhova et al., 2000) (Fig. 3.8). Indeed, for UGA and UAA stop codons, a cytosine at the position +4 seems to provide the best rate of readthrough by most of aminoglycosides, unlike for the UAG stop codon for which an uridine at the position +4 promotes the highest rate of readthrough (Manuvakhova et al., 2000). It is worth noting that all these conclusions were obtained from in vitro or ex vivo analysis and the efficiency of readthrough could be very different in vivo, as already reported (Manuvakhova et al., 2000).

Two types of molecules have been used to induce PTC-readthrough, such as suppressor transfer RNAs (tRNAs) or chemical molecules. The first category is done by changing the anticodon sequence of a tRNA so that it has to recognize one of the stop codon (Fig. 3.9). Since this tRNA still carries an amino acid, it will change a stop codon into a sense codon.

Although suppressor tRNAs have been mainly studied in inferior eukaryotes, natural suppressor tRNAs have also been identified in mammals (Kuchino and Muramatsu, 1996). Interestingly, such natural suppressor

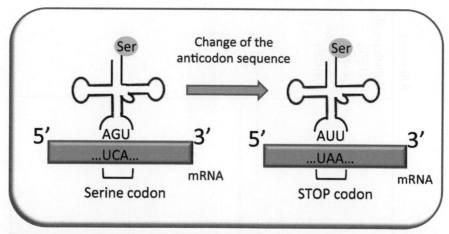

Figure 3.9 *Transformation of a tRNA into a suppressor tRNA by changing the anticodon UGA of the tRNA carrying the serine amino acid into the anticodon UUA to recognize the stop codon UAA.*

tRNAs are tightly regulated in order to act on specific stop codon and to respond to particular stimuli at the accurate moment. In the case of selenoproteins, for example, a tRNA carrying a serine is incorporated at the position of a UGA stop codon. The readthrough of the UGA stop codon is conditioned by the presence of selenium, and is highly regulated via a *cis* sequence on the mRNA encoding the selenoprotein, called SECIS (for selenocysteine insertion sequence) that adopts a specific kink-turn secondary structure, and at least three proteins (the SECIS binding protein 2 (SBP2), the selenocysteine-specific translation elongation factor (eEFSec), and the ribosomal protein L30) (Chavatte et al., 2005; Copeland et al., 2000; Fagegaltier et al., 2000; Tujebajeva et al., 2000). Due to a methylation on one uridine from the anticodon (UUG→UmUG), tRNAs carrying glutamine amino acid have also been shown to read the UAG stop codon, in addition to the CAA codon (Kuchino and Muramatsu, 1996). By gaining a deeper understanding of the regulation of these natural suppressor tRNAs, the use of such particular tRNAs could represent an additional therapeutic strategy to correct nonsense mutations (Temple et al., 1982).

3.1.1 Aminoglycoside Molecules

Chemical molecules represent the second type of molecules capable of readthrough. Indeed, some particular chemical molecules have been shown to improve the efficiency of PTC-readthrough (Table 3.2). Some aminoglycoside family members for instance have the capacity to efficiently readthrough PTC with no effects on the physiological stop codons. Aminoglycosides are

Table 3.2 Molecules with readthrough activity

Family drug	Readthrough drug	Drug structure	References
Aminoglycoside compounds	Gentamicin	Gentamicin C1; Gentamicin C2; Gentamicin C1a	Wilhelm et al. (1978); Burke and Mogg (1985)
	Geneticin (G418)		
	Paromomycin		

		Mattis et al. (2006)
Amikacin		
Tobramycin		
Pyranamycins and kanamycin B analogs	Pyranmycin: "TC derivatives"	 TC001 TC003 TC007 TC032

(Continued)

Table 3.2 Molecules with readthrough activity (*cont.*)

Family drug	Readthrough drug	Drug structure		References
		Kanamycin: "JL derivatives"		Nudelman et al. (2006, 2009, 2010); Shalev et al. (2013)
		JL022	JL023	
	Paromomycin derivatives	NB30	NB54	
		NB74	NB84	

Nonamino-glycoside compounds	Ataluren (PTC124)		Welch et al. (2007)
	Amlexanox		Gonzalez–Hilarion et al. (2012)
	Low molecular mass readthrough compounds (RTC)		Du et al. (2009)

(Continued)

Table 3.2 Molecules with readthrough activity (cont.)

Family drug	Readthrough drug	Drug structure	References
	Acetyl amino benzoic acid compounds	3-{2-(4-Isopropyl-3-methyl-phenoxy)-acetylamino}-benzoic acid 3-[2-4-{1, 1-Dimethyl-propyl)-phenoxy] acetylamino}-benzoic acid	Wilde et al. (2007a)
		3-{2-(4-tert-Butyl-phenoxy)-acetylamino}-benzoic acid	
	Clitocine		Wilde et al. (2007b)

Macrolides	Zilberberg et al. (2010)
Tylosin	
Spiramycin	
Josamycin	
Negamycin	Arakawa et al. (2003a,b)

Source: Arakawa et al. (2003a,b), Burke and Mogg (1985), Mattis et al. (2006), Nudelman et al. (2006, 2009, 2010), Shalev et al. (2013), Wilhelm et al. (1978), Zilberberg et al. (2010).

composed by a sugar molecule carrying at least one amino substitution. They were identified for their antibiotic property at the origin. Among the members of this family, kanamycin, tobramycin, neomycin, or amikacin have been studied for several decades. However, gentamicin and geneticin (also called G418) have been studied deeply for their capacity of correcting PTCs. The efficiency of readthrough is variable, according to the aminoglycoside molecule and, in most cases, does not exceed 5% of the wild-type expression, with molecules such as geneticin or negamycin (Allamand et al., 2008). By comparing the different aminoglycoside members for their capacity to induce readthrough, it appears that a hydroxyl at the position C6' in the ring I is present in aminoglycosides, with efficient readthrough activity suggesting that this chemical group could play an essential role in the mechanism of readthrough (Manuvakhova et al., 2000) (Table 3.2). Aminoglycosides have been tested in mouse models and in human, and showed successfully the rescue of the expression of PTC-containing genes (Du et al., 2002; Guerin et al., 2008; Rowe et al., 2011; Wang et al., 2012; Xue et al., 2014). They were also shown to restore successfully the expression of CFTR in human clinical trials (Sermet-Gaudelus et al., 2007; Wilschanski et al., 2003) but it has been known for a long time that the use of aminoglycosides for a long period of time leads to irreversible ototoxicity and reversible nephrotoxicity (Greenwood, 1959; Hettig and Adcock, 1946; Hinshaw et al., 1946; Hock and Anderson, 1995; Swan, 1997; Toubeau et al., 1986). About 25% of patients treated with aminoglycosides developed a form of ototoxicity. According to the aminoglycoside, the ototoxicity is predominantly cochleotoxic with kanamycin, neomycin, or amikacin or vestibulotoxic with gentamicin or tobramycin (Matz, 1993). The ototoxicity appears few days after systemic exposure to aminoglycosides (Heck et al., 1963). Unlike what was originally hypothesized, the toxicity of aminoglycosides is unrelated to their ability of readthrough nonsense codons, and would rather be related to their capacity of inhibiting translation from mitochondrial ribosomes (Shulman et al., 2014). Indeed, it has been demonstrated that aminoglycosides interact with mitochondrial ribosomes (Qian and Guan, 2009).

Concerning the toxic aspect of aminoglycosides, three different ways have been explored in order to decrease it and to continue with the readthrough therapeutic approach using aminoglycosides or other molecules. The first one is to block the toxicity of aminoglycosides by combining their effect with some drugs that decrease the toxicity of aminoglycosides. The second is to chemically modify aminoglycoside members in order to decrease their toxic effect without affecting their readthrough efficiency, or by increasing it (Nudelman et al., 2006). The third solution is to forget about aminoglycosides and look for readthrough molecules unrelated to the aminoglycoside family.

One possibility to decrease the toxicity of aminoglycosides is to do a cotreatment with another molecule dedicated to blocking the toxicity of aminoglycosides. Aminoglycoside toxicity appears by an increase in inner ear cell death by apoptosis and the production of reactive oxygen species (ROS). Several combinations of treatment have been explored in order to inhibit these consequences. Indeed, antiapoptotic molecules, such as the zVAD-fmk, a pan inhibitor of caspases, has been shown to prevent apoptotic damages by aminoglycosides in vitro, as well as in vivo (Cheng et al., 2003; Forge and Li, 2000; Matsui et al., 2003; Nakagawa et al., 2003; Okuda et al., 2005; Williams and Holder, 2000). Blocking the function of kinases involved in apoptosis activation, such as c-Jun N-terminal kinase (JNK) also shows some efficient results in vitro and in vivo, as well in particular under neomycin exposure (Bodmer et al., 2002; Bonny et al., 2001; Matsui et al., 2004; Nakamagoe et al., 2010; Sugahara et al., 2006; Wang et al., 2003; Ylikoski et al., 2002). Antiapoptotic treatment can only be a short-term solution, due to the risk of inducing a tumorigenesis process when apoptosis is inhibited. Unfortunately, aminoglycosides are not metabolized, and can be detected in cells for months, suggesting that aminoglycoside toxicity can occur even after an antiapoptotic treatment. Therefore, the antiapoptotic strategy is likely not the best cotreatment to provide to patients that would be exposed to aminoglycosides.

To neutralize the ROS produced during a treatment with aminoglycosides, iron chelators have been tested, such as deferoxamine, 2,3 dihydroxybenzoate, or aspirin (Lecain et al., 2007; Sinswat et al., 2000; Song and Schacht, 1996; Wu et al., 2001). The latter has been tested in clinical trials and shows a real efficiency in the decrease of the ototoxicity of aminoglycosides (Behnoud et al., 2009; Chen et al., 2007; Sha et al., 2006) and could represent a practical solution, since aspirin is a very well-known drug and is very well tolerated.

In parallel to trying to decrease the toxicity of aminoglycosides by a cotreatment with other molecules, some laboratories synthesized several new aminoglycoside-derivative molecules, and tested their readthrough efficiency and toxicity. Among them, three molecules named NB54, NB84, and NB124, show very promising results ex vivo, in culture cells, as well as in vivo, in mouse models (Rowe et al., 2011; Wang et al., 2012; Xue et al., 2014). These molecules are more efficient than gentamicin or geneticin, and less toxic than the original aminoglycosides. Thanks to these structure-function studies, an optimization of aminoglycosides is occurring, and more molecules become available that will hopefully reach the clinical phase trials.

The last option to solve the issue of aminoglycoside toxicity is to look for molecules that are not related to aminoglycosides and are capable of rescuing the expression of PTC-containing genes with high readthrough efficiency, and with a low toxicity (see chapter 3 section 3.1.2).

3.1.2 Nonaminoglycoside Molecules

Due to the toxicity of aminoglycosides, screenings have been developed in order to identify various molecules with the capacity of PTC-readthrough, hoping to find molecules with no or low toxicity (Table 3.2). In 2007, the result of a screening by the company PTC therapeutics has been reported, and at least one molecule has been identified. It was initially named PTC124 and is now known as ataluren (Welch et al., 2007). Like aminoglycosides, PTC124 can readthrough UGA codons more efficiently, then UAG, and modestly, UAA stop codons. The efficiency of PTC124 was demonstrated ex vivo by incubating cells in culture harboring a nonsense mutation with PTC124 or DMSO as a control. Readthrough activity was monitored by western-blot, by measuring the level of the full-length protein encoded by the PTC-containing gene. The full-length protein was clearly observed in the presence of PTC124, validating the results of the screening assay. These results were confirmed in vivo by exposing MDX mice harboring a nonsense mutation in the gene encoding the dystrophin. Authors followed the physical capacities of mice, and compared mice exposed to PTC124, and mice exposed to the mock buffer. Clearly, mice exposed to PTC124 showed an increased physical activity, and the expression of dystrophin in cells was monitored by immunofluorescence, using an antibody raised against dystrophin, and coupled with a fluorophore. Since this original study, PTC124 has been successfully tested on different mouse models, such as cystic fibrosis (Du et al., 2008), Usher syndrome (Goldmann et al., 2011, 2012), DMD (Kayali et al., 2012), aniridia (Gregory-Evans et al., 2014), infantile neuronal ceroid lipofuscinosis (Miller et al., 2015), or isolated proximal renal tubular acidosis (Fang et al., 2015). Consistent with these in vivo results, clinical trials were then started with encouraging results. The molecule is very well tolerated by patients and some improvements were observed in patients with DMD or cystic fibrosis. For instance, patients with DMD included in the clinical phase IIb, and receiving a low dosage of ataluren (10 mg/kg in the morning, 10 mg/kg at midday, and

20 mg/kg in the evening) were able to walk 13 m less after 48 weeks of treatment at the 6-MWT than at the beginning of the clinical trial, when they could walk 360 m. Patients receiving the placebo or a high dosage of ataluren (20 mg/kg in the morning, 20 mg/kg at midday, and 40 mg/kg in the evening) lost about 42 m of walking distance at the 6-MWT, at the end of the clinical trial (Haas et al., 2015). These results demonstrate that ataluren slows down the loss of the physical capacity of DMD patients. That is already a very encouraging result for patients and clinicians, since not treatment to cure DMD is available yet. In the case of cystic fibrosis, positive results in human were quickly obtained with the rescue of the expression of CFTR in patients harboring a nonsense mutation in this gene (Sermet-Gaudelus et al., 2010), and the results of the clinical phase III were released in 2014 (Kerem et al., 2014). After 48 weeks, ataluren improved the forced expiratory volume in 1 s (FEV1) of about 5.7%, compared to patients taking the placebo. These results showed an encouraging improvement of the pulmonary function, confirmed by a decrease in the number of exacerbations for the group of patients taking ataluren. This study also highlighted interference between ataluren and tobramycin, an aminoglycoside used as antibiotic, in particular in pulmonary infection, as it is the case for cystic fibrosis patients. Ataluren is the first molecule dedicated to correct nonsense mutations by readthrough that reaches clinical phase III, providing a real hope for patients with genetic diseases caused by nonsense mutations. The results of clinical trials remain modest, but they are encouraging, and especially because no other drugs with a higher efficacy and no toxicity are available yet.

The clinical results of ataluren stimulated the scientific and medical community to find other molecules capable of correcting nonsense mutations efficiently. To date, several other nonaminoglycosides readthrough molecules have been identified and published. Among them, two molecules called readthrough compounds (RTC) 13 and 14 have been shown to induce readthrough very efficiently on ATM, or dystrophin gene harboring a nonsense mutation (Du et al., 2009). The efficiency of both compounds was compared to gentamicin or geneticin, and showed a very comparable effect on nonsense mutation readthrough. In addition, the comparison with ataluren shows that RTC13 or RTC14 can be even more efficient than ataluren on some of the tested models (Kayali et al., 2012; Kuschal et al., 2013). The in vivo demonstration of the efficacy of RTC13, and the absence of toxicity observed in mouse suggests that this compound could represent another possibility for patients with a nonsense mutation-related disease.

3.2 Weaknesses

PTC readthrough is a very attractive approach to correct nonsense mutations because the result is the synthesis of a full-length protein that has a comfortable chance of being functional, except if the mutant position is essential for the function of the protein and the amino acid incorporated at the PTC position is not compatible with this function. All the quantitative assays focusing on readthrough efficiency reported no more than 5% production of readthrough protein, compared to the level of the wild-type protein in nonmutant cells (Bidou et al., 2004; Kayali et al., 2012; Wang et al., 2012). For some pathologies, such as cystic fibrosis, such a low functional protein is enough to decrease strongly at least the pathological phenotype (Ramalho et al., 2002) but, for most of the diseases, a higher rate of readthrough proteins is necessary in order to restore the cellular function of the mutant protein.

Two main reasons can explain why PTC readthrough has limited efficacy. First, PTC readthrough is a mechanism by which the translation machinery is forced to introduce a mistake. The translation process has a high fidelity rate of about 1 error in 10,000 amino acids (Ibba and Soll, 1999; Loftfield and Vanderjagt, 1972); thanks to quality control mechanisms that protect cells from synthesizing aberrant proteins. The readthrough mechanism has to challenge these translation quality controls, which is a very difficult task due to the requirement of having a high fidelity in translation. The second reason is that mRNA candidates for PTC readthrough are first substrates for NMD, making the amount of mRNAs available for readthrough inexistent or reduced to a very low amount.

Another potential limitation in the use of readthrough as a therapeutic approach is the risk of inducing a readthrough on the normal termination codon. As we saw at the beginning of this chapter, readthrough can occur on normal termination codons (Loughran et al., 2014; Yamaguchi et al., 2012). However, readthrough on normal termination codons happens on specific genes that have likely developed a dedicated environment to facilitate the readthrough. The nucleotide sequence around the stop codon has evolved to favor the readthrough and even though it has not been reported yet, it is expected that some factors are involved in order to regulate the readthrough process on those specific genes. Genes that have been shown to support readthrough on their normal termination codon should be now monitored when readthrough strategy is explored, in order to be able to evaluate some putative side effects.

A general readthrough on all normal termination codon is unlikely. Several lines of evidence can explain why readthrough likely occurs on

PTC, and not on most of normal stop codons. The nucleotide context between a PTC and a physiological stop codon is different. Translation termination has been selected during evolution to occur on the physiological stop codon – that is not the case for a PTC. Indeed, it is thought that the 3' UTR carries some *cis*-regulator elements, such as specific secondary structures or specific binding sites, and some trans regulator elements such as proteins to help the translation termination. For example, the protein TPA1 has been shown to interact with the poly(A) binding protein, and the release factors 1 and 3 bridging the 3' UTR with the translation termination complex (Keeling et al., 2006). The close neighborhood of a PTC likely does not support the same proteins. Recently, the methylation of the nucleotide has been shown to be increased in the 3' UTR, compared to the coding sequence, suggesting that a PTC would be surrounded by a poorly methylated nucleotide area, unlike the physiological stop codon (Meyer et al., 2012). However, the methylation of the PTC area has to be clearly studied in order to exclude the possibility that the methylation status of a PTC-nucleotide environment changes quickly after the introduction of the PTC, in order to become similar to the one found at the normal termination codon. However, prudence has to be maintained, since it has been recently reported that many peptides coming from 3' UTR, that is, the readthrough of physiological termination codons, are found on cell surface after gentamicin treatment of cells. This suggests that we might have overestimated the efficiency of translation termination at the physiological stop codons (Goodenough et al., 2014).

Increasing the efficiency of readthrough by using highly efficient molecules might be dangerous, because the natural fidelity of the translation process could be impacted on, and would generate the synthesis of harmful readthrough proteins abnormally longer, due to the readthrough of the physiological stop codon. In order to increase the amount of readthrough protein without increasing the readthrough on the physiological stop codon, another way has been developed, based on the fact that one reason explaining the low efficiency of readthrough is that PTC-containing mRNAs are first substrates for NMD, before being substrates for readthrough. Therefore, no or a very low amount of RNA is available to become substrate for readthrough. Inhibiting NMD in combination to the activation of readthrough might represent an easier (because molecules have already been identified) and safer way (since the translation process will be mildly affected). The inhibition of NMD will be described in the next chapter and the combination of both strategies will be discussed in Section 9.

4 NMD INHIBITION

4.1 Principle

PTC-containing mRNAs are degraded by NMD, as long as the PTC follows the 50–55 nucleotides rule, and/or the distance between the PABPC1 protein and the PTC position. The consequences of NMD are the silencing of the mutant gene, and generally not the synthesis of a truncated protein. NMD prevents the synthesis of a harmful or a nonfunctional truncated protein (Bhuvanagiri et al., 2010; Brogna and Wen, 2009; Chang et al., 2007; Kervestin and Jacobson, 2012; Popp and Maquat, 2014; Rebbapragada and Lykke-Andersen, 2009). However, the truncated protein that would be synthesized without NMD would sometimes keep a residual or the full function of the wild-type protein. Inhibiting NMD then becomes an obvious therapeutic strategy to correct the consequences of some nonsense mutations.

4.2 Examples

Although NMD inhibition has not yet been evaluated for therapeutic approach, many molecules have been shown to inhibit this process (Table 3.3). Among these NMD inhibitors are found translation inhibitors, such as cycloheximide, emetine, or puromycin, due to the fact that NMD requires translation (Shoemaker and Green, 2012). These molecules are an indirect way to block NMD and would be difficult to use for therapeutic approaches, since they are primarily translation inhibitors.

The first molecules capable of inhibiting NMD and not directly linked to translation inhibition were reported in 2004 (Usuki et al., 2004). In this study, the authors showed that wortmannin or caffeine can inhibit NMD by blocking the function of SMG1, the kinase that phosphorylates UPF1. UPF1 becoming hypophosphorylated, the processing of NMD is stopped and PTC-containing mRNAs are stabilized. Both molecules were tested on Ullrich disease patient cells carrying a frameshift mutation of the collagen VI-α2 gene resulting in the introduction of a PTC in the exon 22 that elicits NMD. The incubation of these cells with 10 μM of wortmannin or 7.5 mM of caffeine leads to two- to threefold increase in collagen VI-α2 mRNA. Under these treatments, the truncated protein was also detected, indicating that the stabilized PTC-containing mRNAs support translation.

In 2007, another molecule has been identified as an NMD inhibitor with no other interfering properties. This molecule called NMD inhibitor 1 (NMDI 1) is an indole derivative molecule. This molecule has been successfully tested in cell cultures, and in vivo in a mouse model harboring a nonsense mutation

Table 3.3 Main NMD inhibitors identified to date

Drug name	Structure	References
Amlexanox		Gonzalez-Hilarion et al. (2012)
Caffeine		Usuki et al. (2004)
Cycloheximide		Shoemaker and Green (2012)
Emetine		Shoemaker and Green (2012)
NMDI-1		Durand et al. (2007)
NMDI-14		Martin et al. (2014)
Pateamine A		Dang et al. (2009)
Puromycine		Shoemaker and Green (2012)

(*Continued*)

Table 3.3 Main NMD inhibitors identified to date *(cont.)*

Drug name	Structure	References
Wortmannin		Usuki et al. (2004)

in the gene encoding the α-ʟ-iduronidase, and showed a two- to threefold increase in the stabilization of PTC-containing genes (Durand et al., 2007; Keeling et al., 2013). After the demonstration of the in vivo effectiveness of NMDI 1, that molecule could represent a good candidate for incorporation in therapeutic approaches for genetic diseases caused by a PTC, either alone, or in a combination with a readthrough activator molecule, as we will see in Section 9. The mode of action of NMDI 1 has been studied and it showed that NMDI 1 interferes with the interaction between UPF1 and SMG5, leading to the accumulation of hyperphosphorylated UPF1 and the inhibition of NMD (Durand et al., 2007; Yamashita et al., 2001).

In 2009, Pateamine A, the natural compound isolated from a marine sponge, has been shown to inhibit NMD about two- to threefold (Dang et al., 2009). Pateamine A interacts with eIF4AI and II to inhibit translation and interacts also with eIF4AIII to inhibit NMD, independently from the translation inhibition. The protein eIF4AIII is a core component of the exon junction complex (EJC) known to mark the exon–exon junctions after splicing (Le Hir et al., 2000a,b) (see chapter: General Aspects Related to Nonsense Mutations; Section 3.3). Pateamine A stabilizes the interaction between UPF1 and UPF3X (UPF3b), interfering with the dynamic of the NMD process (Dang et al., 2009).

Using a screening dedicated to the identification of NMD inhibitors, a drug available on the market has been identified as an NMD inhibitor (Gonzalez-Hilarion et al., 2012). This molecule called amlexanox is used for the treatment of aphthous ulcers, and for the treatment of some forms of asthma (Bailey et al., 2011; Bell, 2005; Gonsalves et al., 2007). Amlexanox has been shown to inhibit NMD on several cell lines harboring a nonsense mutation in p53, dystrophin, or CFTR gene, since nonsense mutation-containing mRNAs were stabilized two- to fourfold. Because amlexanox is already a drug provided to patients for more than 30 years, there is abundant information relative to its biodistribution and toxicity (Bell, 2005). Overall, one of the positive information is to conclude that an NMD inhibitor can be well tolerated by patients and can be safe.

More recently, a new set of NMD inhibitors have been reported based on a screen to identify molecules capable of preventing the interaction between UPF1 and SMG7 (Martin et al., 2014). For that, the authors screened molecules in silico, based on their structure, in order to identify the ones that could insert in a SMG7 pocket involved in SMG7-UPF1 interaction. The strong advantage of this approach is to select NMD inhibitors based on a specific mode of action. Ten compounds have been identified and validated using this virtual screen. Among these compounds, NMDI 14 stabilizes PTC-containing globin mRNA up to sevenfold, and is active at nanomolar concentrations with a low cellular toxicity, making this molecule a very powerful NMD inhibitor.

4.3 Weaknesses

NMD inhibition is a nonspecific approach since all NMD reactions will be affected by this approach. In particular, all natural substrates of NMD that are normally degraded by NMD might be stabilized with deleterious consequences, if some mRNAs encode toxic proteins. However, according to the efficiency of the NMD inhibition, natural substrates are not always stabilized, while nonsense mutation-containing mRNAs are. For example, ex vivo inhibition of NMD by NMDI 1 or amlexanox did not lead to the upregulation of genes using NMD as a regulation pathway, unlike apoptosis activators that have been recently shown to inhibit NMD (Durand et al., 2007; Gonzalez-Hilarion et al., 2012; Jia et al., 2015; Popp and Maquat, 2015). It is likely the NMD inhibition efficiency that can explain why some NMD inhibitors affect the natural substrates of NMD expression and why some others do not. That means that genes harboring a nonsense mutation are more sensitive to a variation in NMD efficiency than mRNAs using NMD for their regulation. A hypothesis could be that the nucleotide environment of the PTC found on natural substrates of NMD has been optimized during evolution to be detected by the NMD machinery. This is not the case for the nucleotide environment around a nonsense mutation. This means that the choice of NMD inhibitor is determinant, and as for the readthrough approach, the most efficient might not be the best choice for therapy.

5 PSEUDOURIDYLATION AT THE PTC

5.1 Principle

Pseudouridylation is a natural modification occurring on uridine base in order to convert it into a pseudouridine (named 5-ribosyluracil, Ψ) (Fig. 3.10). Different classes of RNAs are subject to pseudouridylation, such

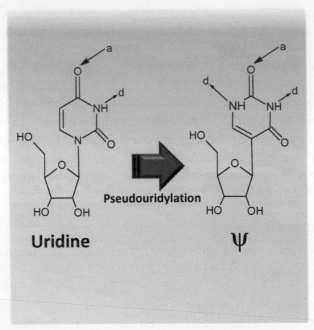

Figure 3.10 *Structural representation of uridine and pseudouridine (Ψ).* Arrows symbolize the acceptor (a) or donor (d) hydrogen bonds established with a complementary nucleotide.

as tRNAs, ribosomal RNAs (rRNAs), and small nuclear RNAs (snRNAs). Pseudouridine has different biochemical properties, compared to uridine; and, in particular, the number of hydrogen bonds that can be established with a complementary nucleotide is different.

The pseudouridylation is endorsed by an H/ACA RNP consisting in a dimer of four proteins named Nhp2p, Nop10p, Gar1p, and Cbf5p, in yeast, and an RNA guide harboring a box H and a box ACA (Fig. 3.11). Interestingly, by modifying the sequence of the RNA guide part, it is possible to decide the uridine to be modified (Huang et al., 2011, 2012). Since all stop codons start by an uridine, they can potentially be subject to pseudouridylation. The conversion of a premature termination codon into a sense codon using pseudouridylation of the first nucleotide of the stop codon has been successfully attempted, in vitro and in vivo in yeast, at least (Karijolich and Yu, 2011). The analysis of the amino acids introduced at the modified stop codon were predominantly the serine or the threonine at the UAA→ΨAA or the UAG→ΨAG targeted stop codon, and the tyrosine or the phenylalanine at the UGA→ΨGA targeted stop codon. This approach is very interesting because by modifying a specific stop codon artificially, it is possible to take advantage of an existing mechanism that will now see a stop codon as a sense codon.

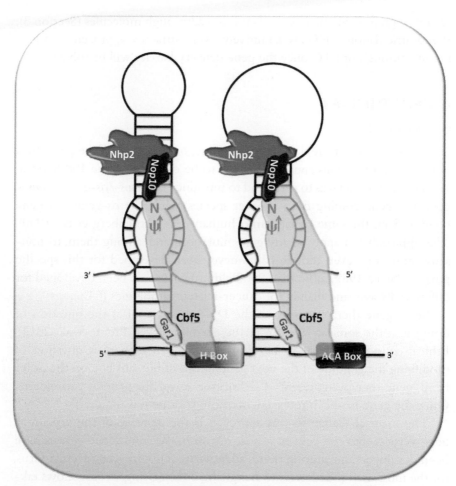

Figure 3.11 *Ribonucleoprotein complex in charge of the pseudouridylation.*

5.2 Weaknesses

As for any targeted method based on the recognition of a specific sequence by annealing, there is always a possibility to anneal at a nonspecific place in the genome. Another limitation in the development of pseudouridylation on a PTC, as a therapeutic approach, is the requirement to introduce in cells a gene expressing the RNA guide or the delivery of an RNA molecule to the cells. The development of this approach is, therefore, related to the development of gene therapy (see Section 6). Finally, the small number of amino acids that can be inserted at the PTC position might reduce the chance of getting functional proteins. However, this limited number of amino acids that can be inserted at the PTC position generates a more homogenous population of readthrough proteins, than with chemical readthrough molecules (Section 3). Pseudouridylation at PTC is definitively a very attractive approach to inducing readthrough of PTC once the gene delivery barrier will be solved.

6 GENE THERAPY

6.1 Principle

Gene therapy aims to replace a mutant gene, or a fraction of the gene, by a wild-type version. This approach started to be explored in the 1980s, using virus such as adenovirus to carry and to introduce the *trans-/cis*-gene (a *trans*-gene is a gene coming from another species, while the *cis*-gene is a gene coming from the same species) into human cells (Rosenberg et al., 1990). This approach can apply to any gene mutations and, among them, to nonsense mutations, even though it has never been developed for this specific purpose. Since 1989, more than 2000 clinical trials have been developed for different diseases, and mainly for cancer-related pathologies (64%).

Since gene therapy occurs at the DNA level, it raises the question of doing it at the somatic level, or at the germline level (Freire et al., 2014). Ethical consideration has to be clearly solved when there is question of modifying the genome of the next generations of humans but, on the other hand, gene therapy represents a definitive answer for genetic pathologies. Once the gene has been replaced, no more treatment is required.

The main challenge to this approach is the delivery of the wild-type gene version into the cells of a host organism. Recombinant viruses have been developed and, among them, adenovirus, retrovirus, or vaccinia virus, for the most used (Niederer and Bangham, 2014). Using viruses allows taking advantage of their systems of delivering a *trans*-gene into the nucleus of a host cell. Some viruses are episomal, so they do not integrate their genome into the host DNA, such as adenovirus, while others integrate their genome in the host DNA, like retrovirus. The consequences can be very different because it is expected that the *trans*-gene brought by adenovirus will be diluted along the multiple cell divisions, while the *trans*-gene brought by retrovirus should be kept in cells and transmitted to daughter cells. However, the integration of virus genome is not stable and it can move from one position to another in the host genome, with the possibility of taking some surrounding DNA at each transfer, raising the risk to inhibit its expression. Similarly, on each integration of the virus genome, the expression of the genes in the vicinity might be disturbed, since the virus genome carries some transcription regulator elements that could increase abnormally some host gene or, in contrast, the integration of the virus genome could destroy a transcriptional element for the host genes.

This therapeutic approach could be used to treat some specific tissues, due to the selectivity of viruses. Indeed, by using specific viruses, it is possible to target preferentially some cell types. For example, adenoviruses have been used to target liver cells (Connelly, 1999), while Herpes simplex virus (HSV) will primarily infect neuron cells (Jacobs et al., 1999). It is definitively a huge advantage of this approach, since the targeting can be very specific and the efficiency of introducing a foreign DNA molecule into human cells is very high.

6.2 Weaknesses

Although this approach is very attractive, because it could treat any types of mutations occurring in a gene, several limitations have been highlighted. The use of virus to introduce the *trans-/cis*-gene, and its integration in the genome of the host cells, raise a question about the consequences of this integration, and the stability of the exonucleic-acid sequence to stay at the integration point. The integration of the genomic sequence of the virus could interfere with the expression of one or several genes by inserting in gene sequences, or by interrupting some regulator elements, such as promoter sequences. The risk associated with the instability of the integration is to see a part of the host genomic sequence being taken away with the genome of the virus. Such event could be at the origin of tumorigenesis, for example, but not only; it could also be at the origin of another genetic disorder by interfering with the expression of a gene related with a pathology. The next challenge will be to target the integration of the *trans-/cis*-gene at a specific neutral locus. Some studies already focus on the targeting of the exogene at a specific position in the host genome, using adenovirus or retrovirus-derivative vectors (Niederer and Bangham, 2014). Although such approach is still fighting with the exclusive integration to the predicted site in the genome, once that issue will be solved, this approach might become the new way to treat any genetic diseases, due to the fact that once the gene is introduced, no more treatment will be necessary, thus decreasing the risk of side effects linked to a long time treatment.

7 CELL THERAPY

7.1 Principle

The injection of cells inside an organism in order to correct a gene expression or to bring a new function is called cell therapy (Fig. 3.12). With the

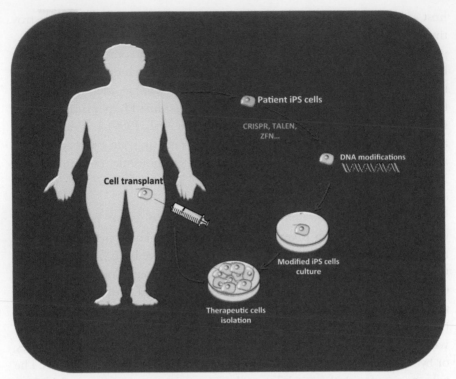

Figure 3.12 *Principle of cell therapy.* Cells are collected from the patient or from a healthy compatible donor and then cultured in vitro to generate IPS or isolate stem cells. At that step, gene therapy or genome editing could also be applied to correct a gene mutation. Finally, the cells can be re-implanted into the patient, where they will be at the origin of a wild-type cell lineage.

emergence of stem cell and induced pluripotent stem (IPS) cell research, cell therapy becomes a very promising therapeutic axis. In the case of nonsense mutation correction, cell therapy would consist in injecting cells with the wild-type version of the mutated gene in the patient. Such approach has not yet been evaluated for the correction of nonsense mutation specifically, but it is adapted in theory. Thanks to the stem cell studies, it is possible to introduce wild-type progenitor cells from several cell lineages that, under external stimuli, will be able to differentiate into one specific lineage, and to maintain the undifferentiated stock of cells. The consequences are a mosaic expression of the targeted gene, with a mutant expression from the original patient cells, and a wild-type expression from cells derived from the injected stem cells. The earlier the cell injection takes places, the more wild-type cells will be present in the adult organism.

Cell therapy always raised ethical questions, and in particular when there appears a question about acting at the embryogenesis step in order to prevent a pathology in a future newborn from affected or carrier parents. However, in the case of pathologies affecting exclusively a cell type, such as leukemia or Crohn diseases, for example, cell therapy represents an exceptional therapeutic hope. In particular, IPS cells fit very well such application, since they are pluripotent and not totipotent, meaning they can only differentiate into several cell lineages, but not all. In addition, their culture is now well understood, making their use more accessible.

One of the actual strategies is to use the stem cells from the patient in order to correct the mutation in these cells by gene therapy or genome editing, and to inject them back into the patient (see Sections 6 and 8). These cells will not induce immune response since they come from the patients themselves, and they will provide the wild-type version of the gene to all the daughter cells. To correct the mutation in stem cells or IPS, gene therapy or genome editing approach can be used, illustrating the requirement to combine therapeutic approaches in order to improve future treatments (see Sections 6 and 8).

Several types of stem cells exist and can be isolated, thanks to specific markers that are dependent on the cell lineages that the stem cell will be able to generate. For instance, CD133 is a surface protein expressed in neurons, and can be used as a neural stem cell marker (Sanai et al., 2005). By selecting cells according to a set of markers found in different cell types derived from a common stem cell, it is possible to isolate these stem cells. However, the stem cell population is very low in a tissue, and can reach up to about 1/100,000. It is, therefore, a big challenge to isolate stem cells from a patient, culture them to expand the population, and then correct the mutant gene before injecting the stem cells back into the patient.

In order to solve the issue of low stem cell population, induced pluripotent stem cells (iPSC) have been developed. To obtain these cells, fibroblasts from a patient, for example, are dedifferentiated in order to lose the maximum of their markers of differentiation. These cells, called iPSC, are then capable of replicating and/or entering in a new differentiation way, depending on the cell culture medium components and, in particular, the content in growth hormones (Takahashi et al., 2007; Takahashi and Yamanaka, 2006). To induce the loss of differentiation, cells are transfected or infected, in order to express at least four specific transcription factors, such as C-MYC, SOX2, KLF4, and OCT3/4 (Takahashi et al., 2007), or other combinations of genes, such as OCT4, SOX2, NANOG, and the translation activator

Lin28 (Yu et al., 2007). The ultimate goal would be to be able to collect some fibroblasts from a skin biopsy, dedifferentiate them in order to differentiate them into neurons, muscle cells, or liver cells, for example.

7.2 Weaknesses

Stem cells have been found in all tissues, indicating that any cell type can be produced from these progenitor cells. The idea of correcting the mutation in these cells before reinjecting them in the patient and letting the differentiation occurs under the control of the patient's body is very attractive. The main risk of this strategy happens at the step where the mutation is corrected in the progenitor cells. For that step, the risks are the same as for the gene therapy or genome editing (Sections 6 and 8) and are related to the genomic DNA modification of these cells. Since cell therapy can modify the germline, ethical considerations have to be solved before going further in the development of this strategy. The putative risk of tumorigenesis and the ethical reasons explain why cell therapy just started some clinical phase trials. However, step-by-step cell therapy becomes safer with, for example, the use of alternative vectors to introduce the reprogramming factors in a safer way. Indeed, retrovirus or lentivirus vectors used at the origin of cell therapy are progressively replaced by adenovirus vectors, limiting the integration in the host genome (Stadtfeld et al., 2008) or the Sendai virus, since this RNA virus replicates in the cytoplasm, excluding any integration in the host genome (Fusaki et al., 2009).

8 GENOME EDITING

Tools to modify a targeted DNA sequence in a genome have recently emerged, and their capacities are promising for the editing of genomes ex vivo. With the development of these tools, we enter the age of molecular surgery. Several categories of editing systems with a similar mode of action have been identified. This mode of action always starts by a cleavage of a specific DNA molecule, and then involves the DNA repair machinery to fix the DNA break. The DNA repair mechanisms that are activated are either the homologous recombination or the nonhomologous end joining (NHEJ) (Fig. 3.13). The first mechanism is very useful when an exosequence needs to be introduced at the cleavage site, while the second mechanism will be of interest in order to inactivate a gene by inducing deletions and frameshift mutations.

Mammalian cells cannot easily repair a double strand break. When double strand break occurs, cells have two methods to solve this issue. The first one, called homologous recombination, requires a template that will be

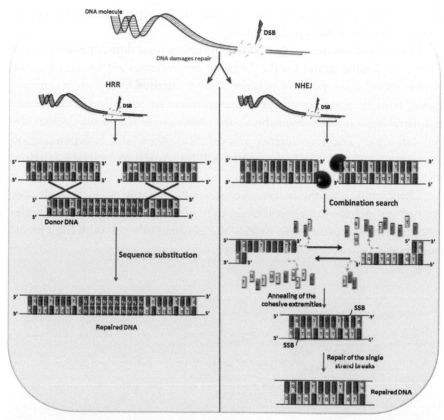

Figure 3.13 *Schematic representation of the homologous recombination repair (HRR) (left panel) and the NHEJ repair (right panel).* HRR involves the replacement of the DNA break area by a donor sequence identical to the region surrounding the DNA break. HRR allows to repair the broken DNA molecule identically as the original one, or to introduce an exogenous sequence at the break position. NHEJ involves first some DNases that degrade DNA from the extremities generated by the DNA break. These DNases have different speed of decay, generating some single strand extremities. Once a partial or a total complementary sequence is found on these single strand extremities, they anneal to generate a DNA molecule with a single strand DNA break on both strands, but not at the same position. Such breaks are easily repaired by base excision repair, nucleotide excision repair, or mismatch repair mechanisms.

integrated at the double strand break position. The homologous recombination allows an exchange of DNA between two DNA molecules surrounded by two identical sequences at each extremity (Fig. 3.13). Repairing the double strand break by homologous recombination raises the risk of getting an insertion of DNA sequence in the gap. This risk becomes a huge advantage when this method will be used for genome editing, since a gene or a part of a gene can be introduced at a specific position. This DNA sequence can be a new protein domain, such as a tag, or the replacement of a missing gene part by the wild-type sequence.

The second method, called NHEJ, illustrates how difficult is for the cell to solve a double strand break. Basically, exonucleases are recruited at the double strand break position in order to degrade the DNA ends at different speed, until they are able to generate cohesive or partially cohesive ends, transforming a double strand break into two single strand breaks that the cell can repair. The consequence is a deletion whose size is depending on the sequence (Fig. 3.13). This method will be favored to generate deletions in a gene by the editing approach.

Three editing methods have recently emerged and, even though they are not yet included in clinical trials, they open new therapeutic strategies by giving the opportunity to modify the genome of a cell with high precision (Fig. 3.14).

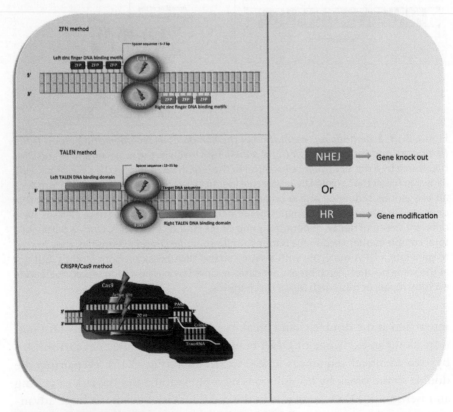

Figure 3.14 *Principle of the three main genome editings.*

8.1 Zinc Finger Nucleases

The zinc finger nucleases (ZFNs) have been reported since the beginning of this century as a tool to induce a double strand break in DNA, at a specific site chosen by the experimenter (Moore et al., 2001). Basically, this tool provides the opportunity to target a fusion protein on any DNA sequence by changing the peptidic sequence of its DNA binding domain (Mani et al., 2005). ZFNs consist in a fusion between a nonspecific endonuclease activity from the C-terminal domain of the restriction enzyme Fok1, and a DNA binding domain composed of at least six zinc fingers capable of recognizing 18 bp (Fig. 3.14). The identity of the amino acids in the zinc fingers dictates the DNA sequence that is recognized. Indeed, by changing the amino acid sequence, it is possible to determine the DNA binding sequence and the cutting site. The probability to find a specific sequence of 18 bp is $1/6.8 \times 10^{10}$, meaning that it would need 22 human genomes to find a given sequence randomly once.

8.1.1 Weaknesses

Several weak points are associated to ZFN technology. The first one is common to any targeting approaches, and is related to specificity, since it cannot be excluded that the ZFN cut DNA at off target sites. The second weak point is shared with gene therapy limitation, since it is about the method of introducing the ZFN or the expression gene encoding the ZFN into the patient cells. Finally, the use of ZFN in patient cells involves the expression of a new protein that can induce an immune response.

8.2 Transcription Activator-Like Effector Nucleases

The transcription activator-like effector nucleases (TALEN) approach is based on the same principle as ZFN, by expressing a fusion protein made of generally the endonuclease FokI to generate the double strand break, and a DNA binding domain derived from transcription activator-like (TAL) proteins, rather than zinc finger domains, as in ZFNs. These proteins isolated from the plant pathogenic *Xanthomonas* (Bai et al., 2000; Kay and Bonas, 2009; White and Yang, 2009; Yang and White, 2004) bind DNA thanks to a central domain consisting of about 34 amino acid tandem repeats. A correlation has been made between the amino acid sequence of the repeats and the nucleotide sequence bound by the TAL (Boch et al., 2009; Moscou and Bogdanove, 2009). In particular, the identity of the amino acids at the position 12 and 13 are hypervariable and are responsible for the amino acid recognition specificity. Increasing or decreasing the number of repeats in a TALEN enzyme affects the specificity of the recognition of the DNA sequence target (Fig. 3.14).

8.2.1 Weaknesses

The same weak points as for ZFN applied to TALEN, since the engineering is similar, and the principle of the DNA sequence recognition is also similar. TALEN have not been included in clinical trials but, as for ZFN, they represent a very useful and precise tool to modify the genome of cells. Some groups reported the use of lentivirus to deliver a TALEN construct in order to perform genome editing in vitro, as well as in vivo, confirming the interest in such technology (Mock et al., 2014).

8.3 CRISPR/Cas9

The system CRISPR/Cas9 is the most recently developed genome editing system (Cong et al., 2013; Jinek et al., 2012; Mali et al., 2013). The principle is different from the two first tools since the recognition of the DNA or RNA sequence is performed via a RNA guide. It is a sequence of about 20 nucleotides that anneals with the target sequence and tethers the nuclease Cas9, since it harbors a high affinity binding sequence for this nuclease. It is therefore very simple to cut a genomic DNA at a specific locus in a host genome by designing a dedicated antisense sequence for the RNA guide. Thanks to the length of the targeted sequence, there is a probability of $1/10^{12}$ to find randomly a specific sequence of 20 nucleotides, meaning such sequence is present once in an equivalent of about 300 human genomes.

The attractive point of this approach is the simplicity of the design for the targeting domain of the enzyme. The RNA guide and the Cas9 can be encoded by two genes present on two different plasmids or on the same one. As for the two precedent genome-editing methods, the synthesis of enzyme is done by the host cell, meaning that the editing of the genome by the CRISPR/Cas9 system only requires a basic transfection of cells.

Like for Fok1 nuclease, the Cas9 cut the two strands of the DNA in a nonspecific sequence. CRISPR/Cas9 will then activate the homologous recombination or the NHEJ repair system (Fig. 3.14). Therefore it is possible to replace a gene or a part of a gene via the CRISPR/Cas9 system, in vitro, at least. It is a very attractive strategy because the cost is reduced and the design of the enzyme is accessible to any labs. This explains the increasing number of reports in less than 3 years concerning this method. Some systems already exist to deliver DNA molecule in specific cells (see Section 6) so they can apply to the CRISPR/Cas9 approach. Some of them have been already tested and, for example, the use of adenovirus expressing CRISPR/Cas9 has been successfully used in vivo (Cheng et al., 2014). Another way to use CRISPR/Cas9 system in therapeutic approaches is to correct a mutation in vitro in patient cells, and inject back the corrected cells into the patient (see Section 7 and the following sections).

8.3.1 Illustration

Although the modification of the genome of an entire complex organism is extremely challenging, CRISPR/Cas9 already demonstrated the potential to be used in a therapeutic development. The author succeeded to reexpress the dystrophin gene harboring frameshift mutations by inducing deletions via CRISPR/Cas9, in order to catch back the wild-type ORF generating an internally truncated dystrophin protein (Ousterout et al., 2015). Authors restored the expression in myoblasts that they injected in mice deficient for dystrophin expression. They observed dystrophin protein in few fibers in mice injected with cells in which dystrophin expression was restored, but not in mice injected with cells in which dystrophin expression was not restored, indicating that the few fibers expressing dystrophin are related to the use of CRISPR/Cas9 and are not revertant cells.

8.3.2 Weaknesses

The two major weak points of the CRISPR/Cas9 approach are the off-targets, and the in vivo delivery, as it is the case for the two other genome-editing methods. Concerning the off-targets, this has already been demonstrated (Cho et al., 2014; Cradick et al., 2013; Fu et al., 2013; Hsu et al., 2013; Ousterout et al., 2015; Pattanayak et al., 2013), so the challenge is to reduce the number of off-targets as much as possible, before exploring the therapeutic advantage of such new technology. For the in vivo delivery concern, genome editing requires to be used with other therapeutic strategy such as cell therapy.

9 COMBINATORY APPROACHES TO IMPROVE NONSENSE MUTATION THERAPIES

Each one of the strategies previously described has strengths and weaknesses that can limit their use as therapeutic approaches. However, some of them can be used together in order to obtain a stronger, new approach. Some of these putative coupled approaches are described next and they only represent few examples.

9.1 Activation of Both Transcription and Readthrough

The low efficiency of PTC-readthrough is often attributed to the low abundance of mRNAs substrates for readthrough, since they are first checked and degraded by NMD before they get a chance to be subject to readthrough.

A way to overcome such issue is to attempt to saturate the NMD mechanism by boosting the transcription. It is likely what is happening in cells expressing CFTR mRNA harboring the W1282X nonsense mutation, under the CMV promoter, and treated with sodium butyrate (Rowe et al., 2007). Sodium butyrate is known to increase the transcription level of genes under CMV promoter and, in this study, the authors succeeded to stabilize very efficiently the PTC-containing CFTR mRNA. As a comparison, they reached about three times the level of wild-type CFTR mRNA present in Calu-3 cells that are already known to overexpress the CFTR gene. The cotreatment of sodium butyrate and geneticin induces a higher presence of CFTR at the cell membrane than with sodium butyrate alone, and a higher CFTR activity than with sodium butyrate or geneticin alone.

9.2 Inhibition of NMD and Activation of Readthrough

Another way to improve PTC-readthrough by increasing the amount of the nonsense mutation-containing mRNA is to inhibit NMD. The efficiency of this approach has been already demonstrated ex vivo by using a siRNA raised against an NMD factor and a gentamicin treatment (Linde et al., 2007). Indeed, in this study, the authors showed that the efficiency of readthrough by gentamicin was improved in the presence of siRNA UPF1 or siRNA UPF2 that induce NMD inhibition, but not in the presence of a nonspecific siRNA. A similar demonstration has been made in vivo using the NMD inhibitor NMDI 1 (Durand et al., 2007), and the readthrough molecule gentamicin in a mouse model harboring the nonsense mutation W392X in the α-L-iduronidase (Keeling et al., 2013).

Some molecules have the dual property to inhibit NMD and promote readthrough. For example, geneticin (G418) is a well-known aminoglycoside with a high efficiency of readthrough (Bidou et al., 2004; Dranchak et al., 2011; Sangkuhl et al., 2004). In addition, geneticin is also an inhibitor of NMD, as it has been demonstrated on DHCR7 mRNA harboring a Q98X or a W151X nonsense mutation, at least (Correa-Cerro et al., 2005). More recently, amlexanox has been selected in a screen as an NMD inhibitor, and it was demonstrated to not only inhibit NMD, but also to promote PTC-readthrough (Gonzalez-Hilarion et al., 2012). Amlexanox shows a higher restoration of the CFTR function in cells harboring a Q2X nonsense mutation in CFTR gene than PTC124/ataluren molecule, for instance. It is therefore possible to find molecules with the dual property that seem to be more efficient than molecules only capable of readthrough; such molecules could represent a better solution than pure readthrough molecules for the correction of nonsense mutations.

9.3 Gene Therapy/Genome Editing/Pseudouridylation at the PTC and Cell Therapy

The use of gene therapy, genome editing, or pseudouridylation at the PTC shows the same limitations that is the delivery of the nonsense mutation cure to all or a specific type of the cells of an organism. Rather than attempting to deliver the cure directly, it could be more efficient to provide the treatment to stem cells or IPS cells, and inject these cured cells in the final organism. By this approach, the correction of the nonsense mutation or the PTC would be done in vitro, with all the advantages related to in vitro manipulations, such as the control of the experimental parameters, for instance. Once the stem cells or the IPS cells are treated, they can be injected in the final organism to play their role of replacing differentiated cells and, by so doing, to provide the corrected expression of the mutant gene. There is still a long way needed to study the behavior of the injected stem cells or IPS cells in a complex organism, but there is no doubt that such approach will be explored in a near future, especially because some clinics already propose cell therapy treatment in some countries.

REFERENCES

Aartsma-Rus, A., Janson, A.A., Kaman, W.E., Bremmer-Bout, M., den Dunnen, J.T., Baas, F., van Ommen, G.J., van Deutekom, J.C., 2003. Therapeutic antisense induced exon skipping in cultured muscle cells from six different DMD patients. Hum. Mol. Genet. 12, 907–914.

Aartsma-Rus, A., Kaman, W.E., Bremmer-Bout, M., Janson, A.A., den Dunnen, J.T., van Ommen, G.J., van Deutekom, J.C., 2004. Comparative analysis of antisense oligonucleotide analogs for targeted DMD exon 46 skipping in muscle cells. Gene Ther. 11, 1391–1398.

Allamand, V., Bidou, L., Arakawa, M., Floquet, C., Shiozuka, M., Paturneau-Jouas, M., Gartioux, C., Butler-Browne, G.S., Mouly, V., Rousset, J.P., et al., 2008. Drug-induced readthrough of premature stop codons leads to the stabilization of laminin alpha2 chain mRNA in CMD myotubes. J. Gene Med. 10, 217–224.

Arakawa, M., Shiozuka, M., Nakayama, Y., Hara, T., Hamada, M., Ikeda, D., Takahashi, Y., Sawa, R., Nonomura, Y., Sheykholeslami, K., Kondo, K., Kaga, K., Takeda, S., Matsuda, R., 2003a. Negamycin-therapy in skeletal and cardiac muscles of mdx mice. Basic Appl. Myol. 13, 313–320.

Arakawa, M., Shiozuka, M., Nakayama, Y., Hara, T., Hamada, M., Kondo, S., Ikeda, D., Takahashi, Y., Sawa, R., Nonomura, Y., et al., 2003b. Negamycin restores dystrophin expression in skeletal and cardiac muscles of mdx mice. J. Biochem. 134, 751–758.

Bai, J., Choi, S.H., Ponciano, G., Leung, H., Leach, J.E., 2000. Xanthomonas oryzae pv. oryzae avirulence genes contribute differently and specifically to pathogen aggressiveness. Mol. Plant Microbe Interact. 13, 1322–1329.

Bailey, J., McCarthy, C., Smith, R.F., 2011. Clinical inquiry. What is the most effective way to treat recurrent canker sores? J. Fam. Pract. 60, 621–632.

Barbash, I.M., Cecchini, S., Faranesh, A.Z., Virag, T., Li, L., Yang, Y., Hoyt, R.F., Kornegay, J.N., Bogan, J.R., Garcia, L., et al., 2013. MRI roadmap-guided transendocardial delivery of exon-skipping recombinant adeno-associated virus restores dystrophin expression in a canine model of Duchenne muscular dystrophy. Gene Ther. 20, 274–282.

Behnoud, F., Davoudpur, K., Goodarzi, M.T., 2009. Can aspirin protect or at least attenuate gentamicin ototoxicity in humans? Saudi Med. J. 30, 1165–1169.

Bell, J., 2005. Amlexanox for the treatment of recurrent aphthous ulcers. Clin. Drug Invest. 25, 555–566.

Berget, S.M., Moore, C., Sharp, P.A., 1977. Spliced segments at the 5′ terminus of adenovirus 2 late mRNA. Proc. Natl. Acad. Sci. USA 74, 3171–3175.

Bhuvanagiri, M., Schlitter, A.M., Hentze, M.W., Kulozik, A.E., 2010. NMD: RNA biology meets human genetic medicine. Biochem. J. 430, 365–377.

Bidou, L., Hatin, I., Perez, N., Allamand, V., Panthier, J.J., Rousset, J.P., 2004. Premature stop codons involved in muscular dystrophies show a broad spectrum of readthrough efficiencies in response to gentamicin treatment. Gene Ther. 11, 619–627.

Blanchet, S., Cornu, D., Argentini, M., Namy, O., 2014. New insights into the incorporation of natural suppressor tRNAs at stop codons in Saccharomyces cerevisiae. Nucleic Acids Res. 42, 10061–10072.

Boch, J., Scholze, H., Schornack, S., Landgraf, A., Hahn, S., Kay, S., Lahaye, T., Nickstadt, A., Bonas, U., 2009. Breaking the code of DNA binding specificity of TAL-type III effectors. Science 326, 1509–1512.

Bodmer, D., Brors, D., Bodmer, M., Ryan, A.F., 2002. Rescue of auditory hair cells from ototoxicity by CEP-11 004, an inhibitor of the JNK signaling pathway. Laryngorhinootologie 81, 853–856.

Bonny, C., Oberson, A., Negri, S., Sauser, C., Schorderet, D.F., 2001. Cell-permeable peptide inhibitors of JNK: novel blockers of beta-cell death. Diabetes 50, 77–82.

Brogna, S., Wen, J., 2009. Nonsense-mediated mRNA decay (NMD) mechanisms. Nat. Struct. Mol. Biol. 16, 107–113.

Burke, J.F., Mogg, A.E., 1985. Suppression of a nonsense mutation in mammalian cells in vivo by the aminoglycoside antibiotics G-418 and paromomycin. Nucleic Acids Res. 13, 6265–6272.

Butland, R.J., Pang, J., Gross, E.R., Woodcock, A.A., Geddes, D.M., 1982. Two-, six-, and 12-minute walking tests in respiratory disease. Br. Med. J. (Clin. Res. Ed.) 284, 1607–1608.

Chang, Y.F., Imam, J.S., Wilkinson, M.F., 2007. The nonsense-mediated decay RNA surveillance pathway. Annu. Rev. Biochem. 76, 51–74.

Chavatte, L., Brown, B.A., Driscoll, D.M., 2005. Ribosomal protein L30 is a component of the UGA-selenocysteine recoding machinery in eukaryotes. Nat. Struct. Mol. Biol. 12, 408–416.

Chen, Y., Huang, W.G., Zha, D.J., Qiu, J.H., Wang, J.L., Sha, S.H., Schacht, J., 2007. Aspirin attenuates gentamicin ototoxicity: from the laboratory to the clinic. Hear. Res. 226, 178–182.

Cheng, A.G., Cunningham, L.L., Rubel, E.W., 2003. Hair cell death in the avian basilar papilla: characterization of the in vitro model and caspase activation. J. Assoc. Res. Otolaryngol. 4, 91–105.

Cheng, R., Peng, J., Yan, Y., Cao, P., Wang, J., Qiu, C., Tang, L., Liu, D., Jin, J., Huang, X., et al., 2014. Efficient gene editing in adult mouse livers via adenoviral delivery of CRISPR/Cas9. FEBS Lett. 588, 3954–3958.

Cho, S.W., Kim, S., Kim, Y., Kweon, J., Kim, H.S., Bae, S., Kim, J.S., 2014. Analysis of off-target effects of CRISPR/Cas-derived RNA-guided endonucleases and nickases. Genome Res. 24, 132–141.

Chow, L.T., Gelinas, R.E., Broker, T.R., Roberts, R.J., 1977. An amazing sequence arrangement at the 5′ ends of adenovirus 2 messenger RNA. Cell 12, 1–8.

Cong, L., Ran, F.A., Cox, D., Lin, S., Barretto, R., Habib, N., Hsu, P.D., Wu, X., Jiang, W., Marraffini, L.A., et al., 2013. Multiplex genome engineering using CRISPR/Cas systems. Science 339, 819–823.

Connelly, S., 1999. Adenoviral vectors for liver-directed gene therapy. Curr. Opin. Mol. Ther. 1, 565–572.

Copeland, P.R., Fletcher, J.E., Carlson, B.A., Hatfield, D.L., Driscoll, D.M., 2000. A novel RNA binding protein, SBP2, is required for the translation of mammalian selenoprotein mRNAs. EMBO J. 19, 306–314.

Correa-Cerro, L.S., Wassif, C.A., Waye, J.S., Krakowiak, P.A., Cozma, D., Dobson, N.R., Levin, S.W., Anadiotis, G., Steiner, R.D., Krajewska-Walasek, M., et al., 2005. DHCR7 nonsense mutations and characterisation of mRNA nonsense mediated decay in Smith-Lemli-Opitz syndrome. J. Med. Genet. 42, 350–357.

Cradick, T.J., Fine, E.J., Antico, C.J., Bao, G., 2013. CRISPR/Cas9 systems targeting beta-globin and CCR5 genes have substantial off-target activity. Nucleic Acids Res. 41, 9584–9592.

Dang, Y., Low, W.K., Xu, J., Gehring, N.H., Dietz, H.C., Romo, D., Liu, J.O., 2009. Inhibition of nonsense-mediated mRNA decay by the natural product pateamine A through eukaryotic initiation factor 4AIII. J. Biol. Chem. 284, 23613–23621.

Davis, R.E., Hardwick, C., Tavernier, P., Hodgson, S., Singh, H., 1995. RNA *trans*-splicing in flatworms. Analysis of *trans*-spliced mRNAs and genes in the human parasite, *Schistosoma mansoni*. J. Biol. Chem. 270, 21813–21819.

Davis, S., Propp, S., Freier, S.M., Jones, L.E., Serra, M.J., Kinberger, G., Bhat, B., Swayze, E.E., Bennett, C.F., Esau, C., 2009. Potent inhibition of microRNA in vivo without degradation. Nucleic Acids Res. 37, 70–77.

Dranchak, P.K., Di Pietro, E., Snowden, A., Oesch, N., Braverman, N.E., Steinberg, S.J., Hacia, J.G., 2011. Nonsense suppressor therapies rescue peroxisome lipid metabolism and assembly in cells from patients with specific PEX gene mutations. J. Cell. Biochem. 112, 1250–1258.

Du, M., Jones, J.R., Lanier, J., Keeling, K.M., Lindsey, J.R., Tousson, A., Bebok, Z., Whitsett, J.A., Dey, C.R., Colledge, W.H., et al., 2002. Aminoglycoside suppression of a premature stop mutation in a Cftr−/− mouse carrying a human CFTR-G542X transgene. J. Mol. Med. (Berl.) 80, 595–604.

Du, M., Liu, X., Welch, E.M., Hirawat, S., Peltz, S.W., Bedwell, D.M., 2008. PTC124 is an orally bioavailable compound that promotes suppression of the human CFTR-G542X nonsense allele in a CF mouse model. Proc. Natl. Acad. Sci. USA 105, 2064–2069.

Du, L., Damoiseaux, R., Nahas, S., Gao, K., Hu, H., Pollard, J.M., Goldstine, J., Jung, M.E., Henning, S.M., Bertoni, C., et al., 2009. Nonaminoglycoside compounds induce readthrough of nonsense mutations. J. Exp. Med. 206, 2285–2297.

Durand, S., Cougot, N., Mahuteau-Betzer, F., Nguyen, C.H., Grierson, D.S., Bertrand, E., Tazi, J., Lejeune, F., 2007. Inhibition of nonsense-mediated mRNA decay (NMD) by a new chemical molecule reveals the dynamic of NMD factors in P-bodies. J. Cell. Biol. 178, 1145–1160.

Egholm, M., Buchardt, O., Christensen, L., Behrens, C., Freier, S.M., Driver, D.A., Berg, R.H., Kim, S.K., Norden, B., Nielsen, P.E., 1993. PNA hybridizes to complementary oligonucleotides obeying the Watson-Crick hydrogen-bonding rules. Nature 365, 566–568.

England, S.B., Nicholson, L.V., Johnson, M.A., Forrest, S.M., Love, D.R., Zubrzycka-Gaarn, E.E., Bulman, D.E., Harris, J.B., Davies, K.E., 1990. Very mild muscular dystrophy associated with the deletion of 46% of dystrophin. Nature 343, 180–182.

Fagegaltier, D., Hubert, N., Yamada, K., Mizutani, T., Carbon, P., Krol, A., 2000. Characterization of mSelB, a novel mammalian elongation factor for selenoprotein translation. EMBO J. 19, 4796–4805.

Fang, Y.W., Yang, S.S., Chau, T., Nakamura, M., Yamazaki, O., Seki, G., Yamada, H., Hsu, H.M., Cheng, C.J., Lin, S.H., 2015. Therapeutic effect of prenatal alkalization and PTC124 in Na/HCO cotransporter 1 p.W516* knock-in mice. Gene Ther. 22, 374–381.

Feng, Y.X., Copeland, T.D., Oroszlan, S., Rein, A., Levin, J.G., 1990. Identification of amino acids inserted during suppression of UAA and UGA termination codons at the gag–pol junction of Moloney murine leukemia virus. Proc. Natl. Acad. Sci. USA 87, 8860–8863.

Floquet, C., Hatin, I., Rousset, J.P., Bidou, L., 2012. Statistical analysis of readthrough levels for nonsense mutations in mammalian cells reveals a major determinant of response to gentamicin. PLoS Genet. 8, e1002608.

Flouriot, G., Brand, H., Seraphin, B., Gannon, F., 2002. Natural *trans*-spliced mRNAs are generated from the human estrogen receptor-alpha (hER alpha) gene. J. Biol. Chem. 277, 26244–26251.

Forge, A., Li, L., 2000. Apoptotic death of hair cells in mammalian vestibular sensory epithelia. Hear. Res. 139, 97–115.

Freire, J.E., Medeiros, S.C., Lopes Neto, A.V., Monteiro Junior, J.E., Sousa, A.J., Rocha, A.J., Menezes, L.M., 2014. Bioethical conflicts of gene therapy: a brief critical review. Rev. Assoc. Med. Bras. 60, 520–524.

Fu, Y., Foden, J.A., Khayter, C., Maeder, M.L., Reyon, D., Joung, J.K., Sander, J.D., 2013. High-frequency off-target mutagenesis induced by CRISPR-Cas nucleases in human cells. Nat. Biotechnol. 31, 822–826.

Fusaki, N., Ban, H., Nishiyama, A., Saeki, K., Hasegawa, M., 2009. Efficient induction of transgene-free human pluripotent stem cells using a vector based on Sendai virus, an RNA virus that does not integrate into the host genome. Proc. Jpn. Acad. Ser. B Phys. Biol. Sci. 85, 348–362.

Goemans, N.M., Tulinius, M., van den Akker, J.T., Burm, B.E., Ekhart, P.F., Heuvelmans, N., Holling, T., Janson, A.A., Platenburg, G.J., Sipkens, J.A., et al., 2011. Systemic administration of PRO051 in Duchenne's muscular dystrophy. N. Engl. J. Med. 364, 1513–1522.

Goldmann, T., Overlack, N., Wolfrum, U., Nagel-Wolfrum, K., 2011. PTC124-mediated translational readthrough of a nonsense mutation causing Usher syndrome type 1C. Hum. Gene Ther. 22, 537–547.

Goldmann, T., Overlack, N., Moller, F., Belakhov, V., van Wyk, M., Baasov, T., Wolfrum, U., Nagel-Wolfrum, K., 2012. A comparative evaluation of NB30, NB54 and PTC124 in translational read-through efficacy for treatment of an USH1C nonsense mutation. EMBO Mol. Med. 4, 1186–1199.

Gonsalves, W.C., Chi, A.C., Neville, B.W., 2007. Common oral lesions: part I. Superficial mucosal lesions. Am. Fam. Physician 75, 501–507.

Gonzalez-Hilarion, S., Beghyn, T., Jia, J., Debreuck, N., Berte, G., Mamchaoui, K., Mouly, V., Gruenert, D.C., Deprez, B., Lejeune, F., 2012. Rescue of nonsense mutations by amlexanox in human cells. Orphanet J. Rare Dis. 7, 58.

Goodenough, E., Robinson, T.M., Zook, M.B., Flanigan, K.M., Atkins, J.F., Howard, M.T., Eisenlohr, L.C., 2014. Cryptic MHC class I-binding peptides are revealed by aminoglycoside-induced stop codon read-through into the 3′ UTR. Proc. Natl. Acad. Sci. USA 111, 5670–5675.

Goyenvalle, A., Babbs, A., van Ommen, G.J., Garcia, L., Davies, K.E., 2009. Enhanced exon-skipping induced by U7 snRNA carrying a splicing silencer sequence: promising tool for DMD therapy. Mol. Ther. 17, 1234–1240.

Goyenvalle, A., Babbs, A., Wright, J., Wilkins, V., Powell, D., Garcia, L., Davies, K.E., 2012. Rescue of severely affected dystrophin/utrophin-deficient mice through scAAV-U7snRNA-mediated exon skipping. Hum. Mol. Genet. 21, 2559–2571.

Greenwood, G.J., 1959. Neomycin ototoxicity; report of a case. AMA Arch. Otolaryngol. 69, 390–397.

Gregory-Evans, C.Y., Wang, X., Wasan, K.M., Zhao, J., Metcalfe, A.L., Gregory-Evans, K., 2014. Postnatal manipulation of Pax6 dosage reverses congenital tissue malformation defects. J. Clin. Invest. 124, 111–116.

Guerin, K., Gregory-Evans, C.Y., Hodges, M.D., Moosajee, M., Mackay, D.S., Gregory-Evans, K., Flannery, J.G., 2008. Systemic aminoglycoside treatment in rodent models of retinitis pigmentosa. Exp. Eye Res. 87, 197–207.

Haas, M., Vlcek, V., Balabanov, P., Salmonson, T., Bakchine, S., Markey, G., Weise, M., Schlosser-Weber, G., Brohmann, H., Yerro, C.P., et al., 2015. European Medicines Agency review of ataluren for the treatment of ambulant patients aged 5 years and older with Duchenne muscular dystrophy resulting from a nonsense mutation in the dystrophin gene. Neuromuscul. Disord. 25, 5–13.

Heck, W.E., Hinshaw, H.C., Parsons, H.G., 1963. Auditory ototoxicity in tuberculosis patients treated with a report of the incidence of hearing loss in a series of 1,150 cases. JAMA 186, 18–20.

Hettig, R.A., Adcock, J.D., 1946. Studies on the toxicity of streptomycin for man; a preliminary report. Science 103, 355–357.

Hinshaw, H.C., Feldman, W.H., Pfuetze, K.H., 1946. Treatment of tuberculosis with streptomycin; a summary of observations on one hundred cases. J. Am. Med. Assoc. 132, 778–782.

Hock, R., Anderson, R.J., 1995. Prevention of drug-induced nephrotoxicity in the intensive care unit. J. Crit. Care 10, 33–43.

Hoogaars, W.M., Mouisel, E., Pasternack, A., Hulmi, J.J., Relizani, K., Schuelke, M., Schirwis, E., Garcia, L., Ritvos, O., Ferry, A., et al., 2012. Combined effect of AAV-U7-induced dystrophin exon skipping and soluble activin Type IIB receptor in mdx mice. Hum. Gene Ther. 23, 1269–1279.

Hsu, P.D., Scott, D.A., Weinstein, J.A., Ran, F.A., Konermann, S., Agarwala, V., Li, Y., Fine, E.J., Wu, X., Shalem, O., et al., 2013. DNA targeting specificity of RNA-guided Cas9 nucleases. Nat. Biotechnol. 31, 827–832.

Huang, C., Karijolich, J., Yu, Y.T., 2011. Post-transcriptional modification of RNAs by artificial Box H/ACA and Box C/D RNPs. Methods Mol. Biol. 718, 227–244.

Huang, C., Wu, G., Yu, Y.T., 2012. Inducing nonsense suppression by targeted pseudouridylation. Nat. Protoc. 7, 789–800.

Ibba, M., Soll, D., 1999. Quality control mechanisms during translation. Science 286, 1893–1897.

Jacobs, A., Breakefield, X.O., Fraefel, C., 1999. HSV-1-based vectors for gene therapy of neurological diseases and brain tumors: part I. HSV-1 structure, replication and pathogenesis. Neoplasia 1, 387–401.

Jia, J., Furlan, A., Gonzalez-Hilarion, S., Leroy, C., Gruenert, D.C., Tulasne, D., Lejeune, F., 2015. Caspases shutdown nonsense-mediated mRNA decay during apoptosis. Cell Death Differ. 22, 1754–1763.

Jinek, M., Chylinski, K., Fonfara, I., Hauer, M., Doudna, J.A., Charpentier, E., 2012. A programmable dual-RNA-guided DNA endonuclease in adaptive bacterial immunity. Science 337, 816–821.

Jungreis, I., Lin, M.F., Spokony, R., Chan, C.S., Negre, N., Victorsen, A., White, K.P., Kellis, M., 2011. Evidence of abundant stop codon readthrough in *Drosophila* and other metazoa. Genome Res. 21, 2096–2113.

Karijolich, J., Yu, Y.T., 2011. Converting nonsense codons into sense codons by targeted pseudouridylation. Nature 474, 395–398.

Karras, J.G., McKay, R.A., Dean, N.M., Monia, B.P., 2000. Deletion of individual exons and induction of soluble murine interleukin-5 receptor-alpha chain expression through antisense oligonucleotide-mediated redirection of pre-mRNA splicing. Mol. Pharmacol. 58, 380–387.

Kay, S., Bonas, U., 2009. How *Xanthomonas* type III effectors manipulate the host plant. Curr. Opin. Microbiol. 12, 37–43.

Kayali, R., Ku, J.M., Khitrov, G., Jung, M.E., Prikhodko, O., Bertoni, C., 2012. Read-through compound 13 restores dystrophin expression and improves muscle function in the mdx mouse model for Duchenne muscular dystrophy. Hum. Mol. Genet. 21, 4007–4020.

Keeling, K.M., Salas-Marco, J., Osherovich, L.Z., Bedwell, D.M., 2006. Tpa1p is part of an mRNP complex that influences translation termination, mRNA deadenylation, and mRNA turnover in *Saccharomyces cerevisiae*. Mol. Cell. Biol. 26, 5237–5248.

Keeling, K.M., Wang, D., Dai, Y., Murugesan, S., Chenna, B., Clark, J., Belakhov, V., Kandasamy, J., Velu, S.E., Baasov, T., et al., 2013. Attenuation of nonsense-mediated mRNA decay enhances in vivo nonsense suppression. PLoS One 8, e60478.

Kerem, E., Konstan, M.W., De Boeck, K., Accurso, F.J., Sermet-Gaudelus, I., Wilschanski, M., Elborn, J.S., Melotti, P., Bronsveld, I., Fajac, I., et al., 2014. Ataluren for the treatment of nonsense-mutation cystic fibrosis: a randomised, double-blind, placebo-controlled phase 3 trial. Lancet Respir. Med. 2, 539–547.

Kervestin, S., Jacobson, A., 2012. NMD: a multifaceted response to premature translational termination. Nat. Rev. Mol. Cell Biol. 13, 700–712.

Kuchino, Y., Muramatsu, T., 1996. Nonsense suppression in mammalian cells. Biochimie 78, 1007–1015.

Kuschal, C., DiGiovanna, J.J., Khan, S.G., Gatti, R.A., Kraemer, K.H., 2013. Repair of UV photolesions in xeroderma pigmentosum group C cells induced by translational readthrough of premature termination codons. Proc. Natl. Acad. Sci. USA 110, 19483–19488.

Le Hir, H., Izaurralde, E., Maquat, L.E., Moore, M.J., 2000a. The spliceosome deposits multiple proteins 20–24 nucleotides upstream of mRNA exon–exon junctions. EMBO J. 19, 6860–6869.

Le Hir, H., Moore, M.J., Maquat, L.E., 2000b. Pre-mRNA splicing alters mRNP composition: evidence for stable association of proteins at exon–exon junctions. Genes Dev. 14, 1098–1108.

Lecain, E., Omri, B., Behar-Cohen, F., Tran Ba Huy, P., Crisanti, P., 2007. The role of PKC-zeta in amikacin-induced apoptosis in the cochlea: prevention by aspirin. Apoptosis 12, 333–342.

Linde, L., Boelz, S., Nissim-Rafinia, M., Oren, Y.S., Wilschanski, M., Yaacov, Y., Virgilis, D., Neu-Yilik, G., Kulozik, A.E., Kerem, E., et al., 2007. Nonsense-mediated mRNA decay affects nonsense transcript levels and governs response of cystic fibrosis patients to gentamicin. J. Clin. Invest. 117, 683–692.

Loftfield, R.B., Vanderjagt, D., 1972. The frequency of errors in protein biosynthesis. Biochem. J. 128, 1353–1356.

Lorain, S., Peccate, C., Le Hir, M., Garcia, L., 2010. Exon exchange approach to repair Duchenne dystrophin transcripts. PLoS One 5, e10894.

Lorain, S., Peccate, C., Le Hir, M., Griffith, G., Philippi, S., Precigout, G., Mamchaoui, K., Jollet, A., Voit, T., Garcia, L., 2013. Dystrophin rescue by *trans*-splicing: a strategy for DMD genotypes not eligible for exon skipping approaches. Nucleic Acids Res. 41, 8391–8402.

Loughran, G., Chou, M.Y., Ivanov, I.P., Jungreis, I., Kellis, M., Kiran, A.M., Baranov, P.V., Atkins, J.F., 2014. Evidence of efficient stop codon readthrough in four mammalian genes. Nucleic Acids Res. 42, 8928–8938.

Lu, Q.L., Rabinowitz, A., Chen, Y.C., Yokota, T., Yin, H., Alter, J., Jadoon, A., Bou-Gharios, G., Partridge, T., 2005. Systemic delivery of antisense oligoribonucleotide restores dystrophin expression in body-wide skeletal muscles. Proc. Natl. Acad. Sci. USA 102, 198–203.

Mali, P., Yang, L., Esvelt, K.M., Aach, J., Guell, M., DiCarlo, J.E., Norville, J.E., Church, G.M., 2013. RNA-guided human genome engineering via Cas9. Science 339, 823–826.

Mani, M., Kandavelou, K., Dy, F.J., Durai, S., Chandrasegaran, S., 2005. Design, engineering, and characterization of zinc finger nucleases. Biochem. Biophys. Res. Commun. 335, 447–457.

Manuvakhova, M., Keeling, K., Bedwell, D.M., 2000. Aminoglycoside antibiotics mediate context-dependent suppression of termination codons in a mammalian translation system. RNA 6, 1044–1055.

Martin, L., Grigoryan, A., Wang, D., Wang, J., Breda, L., Rivella, S., Cardozo, T., Gardner, L.B., 2014. Identification and characterization of small molecules that inhibit nonsense-mediated RNA decay and suppress nonsense p53 mutations. Cancer Res. 74, 3104–3113.

Matsui, J.I., Haque, A., Huss, D., Messana, E.P., Alosi, J.A., Roberson, D.W., Cotanche, D.A., Dickman, J.D., Warchol, M.E., 2003. Caspase inhibitors promote vestibular hair cell survival and function after aminoglycoside treatment in vivo. J. Neurosci. 23, 6111–6122.

Matsui, J.I., Gale, J.E., Warchol, M.E., 2004. Critical signaling events during the aminoglycoside-induced death of sensory hair cells in vitro. J. Neurobiol. 61, 250–266.

Mattis, V.B., Rai, R., Wang, J., Chang, C.W., Coady, T., Lorson, C.L., 2006. Novel aminoglycosides increase SMN levels in spinal muscular atrophy fibroblasts. Hum. Genet. 120, 589–601.

Matz, G.J., 1993. Aminoglycoside cochlear ototoxicity. Otolaryngol. Clin. North. Am. 26, 705–712.

Mendell, J.R., Rodino-Klapac, L.R., Sahenk, Z., Roush, K., Bird, L., Lowes, L.P., Alfano, L., Gomez, A.M., Lewis, S., Kota, J., et al., 2013. Eteplirsen for the treatment of Duchenne muscular dystrophy. Ann. Neurol. 74, 637–647.

Meyer, K.D., Saletore, Y., Zumbo, P., Elemento, O., Mason, C.E., Jaffrey, S.R., 2012. Comprehensive analysis of mRNA methylation reveals enrichment in 3′ UTRs and near stop codons. Cell 149, 1635–1646.

Miller, J.N., Kovacs, A.D., Pearce, D.A., 2015. The novel Cln1(R151X) mouse model of infantile neuronal ceroid lipofuscinosis (INCL) for testing nonsense suppression therapy. Hum. Mol. Genet. 24, 185–196.

Mock, U., Riecken, K., Berdien, B., Qasim, W., Chan, E., Cathomen, T., Fehse, B., 2014. Novel lentiviral vectors with mutated reverse transcriptase for mRNA delivery of TALE nucleases. Sci. Rep. 4, 6409.

Monjaret, F., Bourg, N., Suel, L., Roudaut, C., Le Roy, F., Richard, I., Charton, K., 2014. Cis-splicing and translation of the pre-trans splicing molecule combine with efficiency in spliceosome-mediated RNA trans-splicing. Mol. Ther. 22, 1176–1187.

Moore, M., Klug, A., Choo, Y., 2001. Improved DNA binding specificity from polyzinc finger peptides by using strings of two-finger units. Proc. Natl. Acad. Sci. USA 98, 1437–1441.

Mort, M., Ivanov, D., Cooper, D.N., Chuzhanova, N.A., 2008. A meta-analysis of nonsense mutations causing human genetic disease. Hum. Mutat. 29, 1037–1047.

Moscou, M.J., Bogdanove, A.J., 2009. A simple cipher governs DNA recognition by TAL effectors. Science 326, 1501.

Murphy, W.J., Watkins, K.P., Agabian, N., 1986. Identification of a novel Y branch structure as an intermediate in trypanosome mRNA processing: evidence for trans splicing. Cell 47, 517–525.

Murray, S., Ittig, D., Koller, E., Berdeja, A., Chappell, A., Prakash, T.P., Norrbom, M., Swayze, E.E., Leumann, C.J., Seth, P.P., 2012. TricycloDNA-modified oligo-2′-deoxyribonucleotides reduce scavenger receptor B1 mRNA in hepatic and extrahepatic tissues—a comparative study of oligonucleotide length, design and chemistry. Nucleic Acids Res. 40, 6135–6143.

Nakagawa, T., Kim, T.S., Murai, N., Endo, T., Iguchi, F., Tateya, I., Yamamoto, N., Naito, Y., Ito, J., 2003. A novel technique for inducing local inner ear damage. Hear. Res. 176, 122–127.

Nakamagoe, M., Tabuchi, K., Uemaetomari, I., Nishimura, B., Hara, A., 2010. Estradiol protects the cochlea against gentamicin ototoxicity through inhibition of the JNK pathway. Hear. Res. 261, 67–74.

Niederer, H.A., Bangham, C.R., 2014. Integration site and clonal expansion in human chronic retroviral infection and gene therapy. Viruses 6, 4140–4164.

Nudelman, I., Rebibo-Sabbah, A., Shallom-Shezifi, D., Hainrichson, M., Stahl, I., Ben-Yosef, T., Baasov, T., 2006. Redesign of aminoglycosides for treatment of human genetic diseases caused by premature stop mutations. Bioorg. Med. Chem. Lett. 16, 6310–6315.

Nudelman, I., Rebibo-Sabbah, A., Cherniavsky, M., Belakhov, V., Hainrichson, M., Chen, F., Schacht, J., Pilch, D.S., Ben-Yosef, T., Baasov, T., 2009. Development of novel amino-glycoside (NB54) with reduced toxicity and enhanced suppression of disease-causing premature stop mutations. J. Med. Chem. 52, 2836–2845.

Nudelman, I., Glikin, D., Smolkin, B., Hainrichson, M., Belakhov, V., Baasov, T., 2010. Repairing faulty genes by aminoglycosides: development of new derivatives of geneti-cin (G418) with enhanced suppression of diseases-causing nonsense mutations. Bioorg. Med. Chem. 18, 3735–3746.

Okuda, T., Sugahara, K., Takemoto, T., Shimogori, H., Yamashita, H., 2005. Inhibition of cas-pases alleviates gentamicin-induced cochlear damage in guinea pigs. Auris Nasus Larynx 32, 33–37.

Ousterout, D.G., Kabadi, A.M., Thakore, P.I., Majoros, W.H., Reddy, T.E., Gersbach, C.A., 2015. Multiplex CRISPR/Cas9-based genome editing for correction of dystrophin mutations that cause Duchenne muscular dystrophy. Nat. Commun. 6, 6244.

Pattanayak, V., Lin, S., Guilinger, J.P., Ma, E., Doudna, J.A., Liu, D.R., 2013. High-throughput profiling of off-target DNA cleavage reveals RNA-programmed Cas9 nuclease specific-ity. Nat. Biotechnol. 31, 839–843.

Popp, M.W., Maquat, L.E., 2014. The dharma of nonsense-mediated mRNA decay in mam-malian cells. Mol. Cells 37, 1–8.

Popp, M.W., Maquat, L.E., 2015. Attenuation of nonsense-mediated mRNA decay facilitates the response to chemotherapeutics. Nat. Commun. 6, 6632.

Puttaraju, M., DiPasquale, J., Baker, C.C., Mitchell, L.G., Garcia-Blanco, M.A., 2001. Messenger RNA repair and restoration of protein function by spliceosome-mediated RNA trans-splicing. Mol. Ther. 4, 105–114.

Qian, Y., Guan, M.X., 2009. Interaction of aminoglycosides with human mitochondrial 12S rRNA carrying the deafness-associated mutation. Antimicrob. Agents Chemother. 53, 4612–4618.

Ramalho, A.S., Beck, S., Meyer, M., Penque, D., Cutting, G.R., Amaral, M.D., 2002. Five percent of normal cystic fibrosis transmembrane conductance regulator mRNA ame-liorates the severity of pulmonary disease in cystic fibrosis. Am. J. Respir. Cell Mol. Biol. 27, 619–627.

Rebbapragada, I., Lykke-Andersen, J., 2009. Execution of nonsense-mediated mRNA decay: what defines a substrate? Curr. Opin. Cell Biol. 21, 394–402.

Rosenberg, S.A., Aebersold, P., Cornetta, K., Kasid, A., Morgan, R.A., Moen, R., Karson, E.M., Lotze, M.T., Yang, J.C., Topalian, S.L., et al., 1990. Gene transfer into humans—immunotherapy of patients with advanced melanoma, using tumor-infiltrating lympho-cytes modified by retroviral gene transduction. N. Engl. J. Med. 323, 570–578.

Rowe, S.M., Varga, K., Rab, A., Bebok, Z., Byram, K., Li, Y., Sorscher, E.J., Clancy, J.P., 2007. Restoration of W1282X CFTR activity by enhanced expression. Am. J. Respir. Cell Mol. Biol. 37, 347–356.

Rowe, S.M., Sloane, P., Tang, L.P., Backer, K., Mazur, M., Buckley-Lanier, J., Nudelman, I., Belakhov, V., Bebok, Z., Schwiebert, E., et al., 2011. Suppression of CFTR premature termination codons and rescue of CFTR protein and function by the synthetic amino-glycoside NB54. J. Mol. Med. (Berl.) 89, 1149–1161.

Sanai, N., Alvarez-Buylla, A., Berger, M.S., 2005. Neural stem cells and the origin of glio-mas. N. Engl. J. Med. 353, 811–822.

Sangkuhl, K., Schulz, A., Rompler, H., Yun, J., Wess, J., Schoneberg, T., 2004. Aminoglycoside-mediated rescue of a disease-causing nonsense mutation in the V2 vasopressin receptor gene in vitro and in vivo. Hum. Mol. Genet. 13, 893–903.

Sermet-Gaudelus, I., Renouil, M., Fajac, A., Bidou, L., Parbaille, B., Pierrot, S., Davy, N., Bismuth, E., Reinert, P., Lenoir, G., et al., 2007. In vitro prediction of stop-codon suppression by intravenous gentamicin in patients with cystic fibrosis: a pilot study. BMC Med. 5, 5.

Sermet-Gaudelus, I., Boeck, K.D., Casimir, G.J., Vermeulen, F., Leal, T., Mogenet, A., Roussel, D., Fritsch, J., Hanssens, L., Hirawat, S., et al., 2010. Ataluren (PTC124) induces cystic fibrosis transmembrane conductance regulator protein expression and activity in children with nonsense mutation cystic fibrosis. Am. J. Respir. Crit. Care Med. 182, 1262–1272.

Sha, S.H., Qiu, J.H., Schacht, J., 2006. Aspirin to prevent gentamicin-induced hearing loss. N. Engl. J. Med. 354, 1856–1857.

Shalev, M., Kondo, J., Kopelyanskiy, D., Jaffe, C.L., Adir, N., Baasov, T., 2013. Identification of the molecular attributes required for aminoglycoside activity against *Leishmania*. Proc. Natl. Acad. Sci. USA 110, 13333–13338.

Shoemaker, C.J., Green, R., 2012. Translation drives mRNA quality control. Nat. Struct. Mol. Biol. 19, 594–601.

Shulman, E., Belakhov, V., Wei, G., Kendall, A., Meyron-Holtz, E.G., Ben-Shachar, D., Schacht, J., Baasov, T., 2014. Designer aminoglycosides that selectively inhibit cytoplasmic rather than mitochondrial ribosomes show decreased ototoxicity: a strategy for the treatment of genetic diseases. J. Biol. Chem. 289, 2318–2330.

Sinswat, P., Wu, W.J., Sha, S.H., Schacht, J., 2000. Protection from ototoxicity of intraperitoneal gentamicin in guinea pig. Kidney Int. 58, 2525–2532.

Song, B.B., Schacht, J., 1996. Variable efficacy of radical scavengers and iron chelators to attenuate gentamicin ototoxicity in guinea pig in vivo. Hear. Res. 94, 87–93.

Stadtfeld, M., Nagaya, M., Utikal, J., Weir, G., Hochedlinger, K., 2008. Induced pluripotent stem cells generated without viral integration. Science 322, 945–949.

Sugahara, K., Rubel, E.W., Cunningham, L.L., 2006. JNK signaling in neomycin-induced vestibular hair cell death. Hear. Res. 221, 128–135.

Summerton, J., Weller, D., 1997. Morpholino antisense oligomers: design, preparation, and properties. Antisense Nucleic Acid Drug Dev. 7, 187–195.

Swan, S.K., 1997. Aminoglycoside nephrotoxicity. Semin. Nephrol. 17, 27–33.

Takahashi, K., Yamanaka, S., 2006. Induction of pluripotent stem cells from mouse embryonic and adult fibroblast cultures by defined factors. Cell 126, 663–676.

Takahashi, K., Tanabe, K., Ohnuki, M., Narita, M., Ichisaka, T., Tomoda, K., Yamanaka, S., 2007. Induction of pluripotent stem cells from adult human fibroblasts by defined factors. Cell 131, 861–872.

Takeshima, Y., Yagi, M., Wada, H., Ishibashi, K., Nishiyama, A., Kakumoto, M., Sakaeda, T., Saura, R., Okumura, K., Matsuo, M., 2006. Intravenous infusion of an antisense oligonucleotide results in exon skipping in muscle dystrophin mRNA of Duchenne muscular dystrophy. Pediatr. Res. 59, 690–694.

Temple, G.F., Dozy, A.M., Roy, K.L., Kan, Y.W., 1982. Construction of a functional human suppressor tRNA gene: an approach to gene therapy for beta-thalassaemia. Nature 296, 537–540.

Toubeau, G., Laurent, G., Carlier, M.B., Abid, S., Maldague, P., Heuson-Stiennon, J.A., Tulkens, P.M., 1986. Tissue repair in rat kidney cortex after short treatment with aminoglycosides at low doses. A comparative biochemical and morphometric study. Lab. Invest. 54, 385–393.

Tujebajeva, R.M., Copeland, P.R., Xu, X.M., Carlson, B.A., Harney, J.W., Driscoll, D.M., Hatfield, D.L., Berry, M.J., 2000. Decoding apparatus for eukaryotic selenocysteine insertion. EMBO Rep. 1, 158–163.

Usuki, F., Yamashita, A., Higuchi, I., Ohnishi, T., Shiraishi, T., Osame, M., Ohno, S., 2004. Inhibition of nonsense-mediated mRNA decay rescues the phenotype in Ullrich's disease. Ann. Neurol. 55, 740–744.

Vulin, A., Barthelemy, I., Goyenvalle, A., Thibaud, J.L., Beley, C., Griffith, G., Benchaouir, R., le Hir, M., Unterfinger, Y., Lorain, S., et al., 2012. Muscle function recovery in golden retriever muscular dystrophy after AAV1-U7 exon skipping. Mol. Ther. 20, 2120–2133.

Walls, Z.F., Puttaraju, M., Temple, G.F., Gambhir, S.S., 2008. A generalizable strategy for imaging pre-mRNA levels in living subjects using spliceosome-mediated RNA trans-splicing. J. Nucl. Med. 49, 1146–1154.

Wally, V., Murauer, E.M., Bauer, J.W., 2012. Spliceosome-mediated trans-splicing: the therapeutic cut and paste. J. Invest. Dermatol. 132, 1959–1966.

Wang, Z., Burge, C.B., 2008. Splicing regulation: from a parts list of regulatory elements to an integrated splicing code. RNA 14, 802–813.

Wang, J., Van De Water, T.R., Bonny, C., de Ribaupierre, F., Puel, J.L., Zine, A., 2003. A peptide inhibitor of c-Jun N-terminal kinase protects against both aminoglycoside and acoustic trauma-induced auditory hair cell death and hearing loss. J. Neurosci. 23, 8596–8607.

Wang, D., Belakhov, V., Kandasamy, J., Baasov, T., Li, S.C., Li, Y.T., Bedwell, D.M., Keeling, K.M., 2012. The designer aminoglycoside NB84 significantly reduces glycosaminoglycan accumulation associated with MPS I-H in the Idua-W392X mouse. Mol. Genet. Metab. 105, 116–125.

Welch, E.M., Barton, E.R., Zhuo, J., Tomizawa, Y., Friesen, W.J., Trifillis, P., Paushkin, S., Patel, M., Trotta, C.R., Hwang, S., et al., 2007. PTC124 targets genetic disorders caused by nonsense mutations. Nature 447, 87–91.

Wheeler, T.M., Leger, A.J., Pandey, S.K., MacLeod, A.R., Nakamori, M., Cheng, S.H., Wentworth, B.M., Bennett, C.F., Thornton, C.A., 2012. Targeting nuclear RNA for in vivo correction of myotonic dystrophy. Nature 488, 111–115.

White, F.F., Yang, B., 2009. Host and pathogen factors controlling the rice-Xanthomonas oryzae interaction. Plant Physiol. 150, 1677–1686.

Wilde, R.G., Welch, E.M., Takasugi, J.J., Almstead, N.G., Rubenstein, S.M., Beckmann, H., 2007a. Acetylamino Benzoic Acid Compounds and their Use for Nonsense Suppression and the Treatment of Disease, vol. 7247741B2. PTC Therapeutics, Inc.; Amgen Inc., US, USPTO.

Wilde, R.G., Kennedy, P.D., Almstead, N.G., Welch, E.M., Takasugi, J.J., Friesen, W.J., 2007b. Nucleoside Compounds and their Use for Treating Cancer and Diseases Associated with Somatic Mutations, vol. 7291603B2. PTC Therapeutics, Inc., USA, USPTO. United States Patent.

Wilhelm, J.M., Jessop, J.J., Pettitt, S.E., 1978. Aminoglycoside antibiotics and eukaryotic protein synthesis: stimulation of errors in the translation of natural messengers in extracts of cultured human cells. Biochemistry 17, 1149–1153.

Williams, J.A., Holder, N., 2000. Cell turnover in neuromasts of zebrafish larvae. Hear. Res. 143, 171–181.

Wilschanski, M., Yahav, Y., Yaacov, Y., Blau, H., Bentur, L., Rivlin, J., Aviram, M., Bdolah-Abram, T., Bebok, Z., Shushi, L., et al., 2003. Gentamicin-induced correction of CFTR function in patients with cystic fibrosis and CFTR stop mutations. N. Engl. J. Med. 349, 1433–1441.

Witten, J.T., Ule, J., 2011. Understanding splicing regulation through RNA splicing maps. Trends Genet. 27, 89–97.

Wu, W.J., Sha, S.H., McLaren, J.D., Kawamoto, K., Raphael, Y., Schacht, J., 2001. Aminoglycoside ototoxicity in adult CBA, C57BL and BALB mice and the Sprague-Dawley rat. Hear. Res. 158, 165–178.

Xue, X., Mutyam, V., Tang, L., Biswas, S., Du, M., Jackson, L.A., Dai, Y., Belakhov, V., Shalev, M., Chen, F., et al., 2014. Synthetic aminoglycosides efficiently suppress cystic fibrosis transmembrane conductance regulator nonsense mutations and are enhanced by ivacaftor. Am. J. Respir. Cell Mol. Biol. 50, 805–816.

Yagi, M., Takeshima, Y., Surono, A., Takagi, M., Koizumi, M., Matsuo, M., 2004. Chimeric RNA and 2'-O, 4'-C-ethylene-bridged nucleic acids have stronger activity than phosphorothioate oligodeoxynucleotides in induction of exon 19 skipping in dystrophin mRNA. Oligonucleotides 14, 33–40.

Yamada, T., Okaniwa, N., Saneyoshi, H., Ohkubo, A., Seio, K., Nagata, T., Aoki, Y., Takeda, S., Sekine, M., 2011. Synthesis of 2'-O-[2-(N-methylcarbamoyl)ethyl]ribonucleosides using oxa-Michael reaction and chemical and biological properties of oligonucleotide derivatives incorporating these modified ribonucleosides. J. Org. Chem. 76, 3042–3053.

Yamaguchi, Y., Hayashi, A., Campagnoni, C.W., Kimura, A., Inuzuka, T., Baba, H., 2012. L-MPZ, a novel isoform of myelin P0, is produced by stop codon readthrough. J. Biol. Chem. 287, 17765–17776.

Yamashita, A., Ohnishi, T., Kashima, I., Taya, Y., Ohno, S., 2001. Human SMG-1, a novel phosphatidylinositol 3-kinase-related protein kinase, associates with components of the mRNA surveillance complex and is involved in the regulation of nonsense-mediated mRNA decay. Genes Dev. 15, 2215–2228.

Yang, B., White, F.F., 2004. Diverse members of the AvrBs3/PthA family of type III effectors are major virulence determinants in bacterial blight disease of rice. Mol. Plant. Microbe Interact. 17, 1192–1200.

Ylikoski, J., Xing-Qun, L., Virkkala, J., Pirvola, U., 2002. Blockade of c-Jun N-terminal kinase pathway attenuates gentamicin-induced cochlear and vestibular hair cell death. Hear. Res. 166, 33–43.

Yu, J., Vodyanik, M.A., Smuga-Otto, K., Antosiewicz-Bourget, J., Frane, J.L., Tian, S., Nie, J., Jonsdottir, G.A., Ruotti, V., Stewart, R., et al., 2007. Induced pluripotent stem cell lines derived from human somatic cells. Science 318, 1917–1920.

Zilberberg, A., Lahav, L., Rosin-Arbesfeld, R., 2010. Restoration of APC gene function in colorectal cancer cells by aminoglycoside- and macrolide-induced read-through of premature termination codons. Gut 59, 496–507.

CHAPTER 4

Conclusions

Contents

1 SUMMARY ON THE DIFFERENT STRATEGIES AND THEIR RESULTS

A wide panel of new therapeutic strategies is explored to propose treatment to genetic diseases. Some of these strategies are very molecular and target the DNA molecule (gene therapy, genome editing) or the RNA molecule (*trans*-splicing, exon skipping, activation of PTC readthrough, inhibition of NMD, pseudouridylation), while some others are more cellular (cell therapy). Many of them are still at the experimental development stage (NMD inhibition, pseudouridylation, *trans*-splicing, genome editing) and some have already reached the clinical phases (gene therapy, exon skipping, PTC readthrough, or cell therapy).

They can all be applied to treat patients with a nonsense mutation, even though only NMD inhibition and readthrough activation are exclusively dedicated to nonsense mutation correction, and could be called nonsense mutation therapies. This means that many patients with different pathologies and eventually different types of mutations could benefit from these therapeutic approaches. It is important to have several options to treat a mutation, because none of the described strategies are perfect, they all have some weak points. However, some will be more adapted to correct a specific nonsense mutation than others because of the mutation and its position in the gene, for example, the gene itself, its tissue-specific expression, or its level of expression. In addition, developing in parallel several therapeutic approaches with convergent applications increases the possibility to deliver a solution to treat nonsense mutations faster, since strategies do not progress at the same speed, and some meet technical limitations that will slow down their achievement. For example, the gene delivery method might be a strong limitation for several of the previously described strategies, such as gene therapy, pseudouridylation, genome editing, or *trans*-splicing. Therapeutic approaches using chemical molecules have, then, less issues on those particular points and might be able to provide a treatment earlier than other strategies that use more complex biology tools.

The ethical question is also an issue that will slow down the development of cell or gene therapy in particular. This question is really important and involves not only scientists or clinicians, but also politicians and all citizens, explaining why the use of embryonic stem cells is forbidden in Germany and Italy; tightly controlled and restricted in the United States of America or France; and much more accessible in other countries, such as Austria, Poland, or Ireland since no specific laws determine the use and the origin of the stem cells. Therefore, the development of cell therapy will be dependent on the countries.

Overall, results from therapeutic strategies described in this book are very encouraging, since the original objective to be able to reexpress the mutant gene is reached by several of them (*trans*-splicing, PTC-readthrough, gene therapy, exon skipping, or cell therapy), at least in vitro and/or in animal models. Solutions are also in development to improve these individual strategies by evaluating combinations of two of them with some already positive results (NMD inhibition and PTC readthrough). Treatments are not yet on the market, but, with several molecules in clinical phase II or III, it should be not too long to see the first treatment for nonsense mutation-related diseases released on the market.

2 PERSONALIZED/TARGETED MEDICINE VERSUS TRADITIONAL MEDICINE

Medicine increasingly treats the causes of pathology, rather than the symptoms. Personalized and targeted medication contribute to this modern way of thinking in medicine. Without human genetic research programs to identify the mutant genes responsible for a pathology and the human genome–sequencing project, targeted, and personalized medication would not be possible today. It has been a long road to link genetic diseases to pathology and to understand the molecular mechanism behind the development of a disease. The function of some genes responsible for a pathology when mutated still remains not fully understood as, for instance, in neurologic disorders or metabolic diseases for which the genetic link is often more recent than for rare diseases.

Thanks to the support of patient associations in particular, progresses have been made in various aspects of molecular, cellular, and medical biology. The fastest and most significant progresses are often related to the size of the patient association behind a disease, which is also often related to the number of patients affected by that pathology. However, associations of patients with a rare disease succeeded to stimulate some of the scientific community to focus on rare genetic diseases. These pathologies are attractive, because it is possible to link one disorder to one gene, which is simpler than multifactorial pathologies. In theory, the treatment of rare diseases is reduced to bring back the missing function encoded by the mutant gene. The basic concept is to understand the function of the gene and the molecular mechanism that is impaired due to the mutation. Many different approaches have been designed as described in this book, focusing on different aspects of the diseases or on the element to replace: the mutant gene, the mRNA harboring the mutation, or the pathologic cell. Importantly, all these therapeutic approaches can apply to any genetic diseases and are not restricted to one pathology. Progress made in the rare disease area might be of interest for more frequent diseases such as cancer.

Thanks to the development of all these therapeutic approaches; soon, patients sharing the same pathology will not receive the same treatment. This will be in order to cure the cause of the pathology, rather than the consequences. In addition, patients with different pathologies might receive the same treatment if the mutation at the origin of the pathology belongs to the same category and can be treated in a similar way, molecularly speaking. A treatment will not be dedicated to one pathology but to one type of mutation, which is really a new way of thinking.

3 LIMITATIONS ON NONSENSE MUTATION THERAPIES AND FUTURE CONSIDERATIONS

The nonsense mutation therapies exclusively dedicated to nonsense mutation correction (inhibition of NMD and activation of readthrough) target the mutant mRNA and do not affect the patient genome. This is certainly a very attractive ethic advantage since the patient will not become a genetically modified organism and the next generation will not be affected. However, such approach is also a potential problem after several generations. With these approaches, the mutation in the gene is not corrected, since the treatment focuses only on the mutant mRNA, making the mutation stable in the population of healthy people. The consequence is that nonsense mutations will be stabilized in the human population and might progressively conquer the human DNA patrimony, if treatment is provided for several generations. Therefore nonsense mutation therapies will have to be considered as a temporary treatment limited to several generations, until other therapies focusing on gene repair become available. Nonsense mutation therapies have to be developed because they are the ones that progress the fastest, meaning that they will be available before other options could be proposed to patients.

Another consideration that has to be kept in mind with nonsense mutation therapies is that both are related to the use of chemical molecules. We know that the treatment will have to be taken for the entire life of the patient. Besides the side effects due to a very long period of exposure, such long time treatment often becomes less efficient after a while. The decrease of treatment efficiency is related to an adaptation of the human body to the presence of the drug by developing some mechanisms to block the entrance of the molecule in cells, exclude it from cells, or metabolize it in order to transform it into an inactive molecule. These mechanisms are called drug resistance mechanisms and are well-known phenomena of long time treatments, in particular for anticancer treatments or antivirus treatments (Clavel and Hance, 2004). This means that several molecules with similar efficiency for the nonsense mutation correction, but with a different mode of action have to be available in order to be able to change the drug when it is less efficient. This issue has already been anticipated, since besides ataluren, several other molecules are being studied (amlexanox, RTC13, RTC14 molecules). This is also the reason why the mode of action of each molecule has to be determined, in order to ensure that the target protein is different for each drug.

Although some treatments to correct nonsense mutations reach the final step before being released on the market, the identification and the characterization of new molecules and treatments are still needed. For that, new technologies, new approaches, and new targets have to be developed and studied, in order to be able to propose a wide panel of solutions adapted to each patient.

REFERENCE

Clavel, F., Hance, A.J., 2004. HIV drug resistance. N. Engl. J. Med. 350, 1023–1035.

GLOSSARY

De novo mutation New mutation present in a family member and absent in parents. Such mutation appears either in one of the germ cells of parents or in the fertilized egg at the origin of the family member carrier of the novo mutation.

Frameshift mutation Mutation inducing a change of the reading frame. This shift can be towards the 3′ end of the mRNA, in which case it is a positive frameshift. When the shift is towards the 5′ end of the mRNA, it is a negative frameshift. The frameshift can be minus or plus 1, indicating that it corresponds to an insertion or a deletion of a multiple of three plus 1 nucleotide. A frameshift plus or minus 2 corresponds to an insertion or a deletion of a multiple of three plus 2 nucleotides.

Homologous Two genes are homologous when they share an ancestral gene at the origin of their evolution.

Missense mutation Mutation leading to the change of one amino acid by another one.

Nonsense mutation Mutation changing a translatable codon into an untranslatable codon called nonsense codon or stop codon, interrupting the open reading frame.

Orthologous Two genes are orthologous when they are homologous and present in two different organisms. For example, UPF1 gene in yeast is an ortholog of UPF1 gene in human.

Paralogous Two genes are paralogous when they are homologous and present in the same organism. For example, in mammalian cells, UPF3/UPF3a gene is a paralog of UPF3X/UPF3b.

Personalized therapy Treatment designed for a unique patient according to the molecular origin of the pathology and his/her genetic and biologic backgrounds.

Targeted therapy Treatment designed for some but not all patients affected by the same pathology or not. Targeted therapies are designed according to the molecular mechanism and/or the type of mutation promoting the pathology. Patients who benefit from a targeted therapy share the same molecular causes at the origin of the pathology.

Transition A transition is a mutation changing a purine base (guanosine or adenosine) into a purine base, or a pyrimidine base (cytosine or thymidine) into a pyrimidine base.

Transversion A transversion is a mutation changing a purine base (guanosine or adenosine) into a pyrimidine base (cytosine or thymidine), or a pyrimidine base into a purine base.

SUBJECT INDEX

Printed and bound by CPI Group (UK) Ltd, Croydon, CR0 4YY

03/10/2024

01040420-0020